Biomimetic Sensor Technology

KIYOSHI TOKO

T0185285

This book deals with the development of what are known as the electronic nose and the electronic tongue, that is to say odor and taste sensors. Of all sensor technologies, these have been widely considered as the most difficult to realize.

These sensors have been developed on the basis of mechanisms found in biological systems, such as parallel processing of multidimensional information or by the use of biomaterials. The author begins with an introductory section dealing with the principles of measurement and multivariate analysis. Reception mechanisms in biological systems are briefly reviewed. It is shown that a taste sensor using lipid membranes can reproduce a sense of taste and provide clues for the reception mechanism in gustatory systems. Several types of biosensor, including enzyme-immobilized membranes, SPR, the quartz resonance oscillator and IC technologies are explained in detail. Generally a sensor is designed to reproduce or surpass the human senses. For this purpose, it is not enough simply to detect a single input signal. Instead, it is necessary to be able to perform adequate processing of diverse kinds of information. The expansion of sensor technologies into everyday life will require the fusion of sensor technologies in various fields. This book is dedicated to the development of intelligent sensors and systems.

This book will be of value to researchers from a wide range of fields, including electrical engineering, food science, electronics, mechanical engineering, medical science, biochemistry, biophysics and biotechnology.

Dr Toko is a Professor of the Department of Electronic Device Engineering, Kyushu University. He graduated from the electronics course at Kyushu University and then in his PhD graduate course started research in biology related to biological and artificial membranes. This involved cooperating with several groups in varied scientific fields, such as food technology, plant physiology, applied chemistry and molecular biology. He received his PhD from Kyushu University in the study of self-organization in biomembranes and biological systems. He continued this work during a period as Research Associate and Associate Professor in the same laboratory. During that time he succeeded in developing the first-ever taste sensor using lipid membranes, i.e. the electronic tongue. At present, this electronic tongue is sold commercially in Japan and is beginning to be distributed all over the world. He is now one of the leading scientists in the field of bioelectronics, which deals with devices and phenomena related to both electronics and biology. He has published more than 300 papers in well-respected journals on the subject of taste sensors and the application of lipid membranes. He directs several government projects in food research using biosensors and the taste sensor. His present concern is to clarify the reception mechanism of the gustatory system based on results obtained using the taste sensor. He is a member of professional associations of applied physics, taste and smell, physics, membrane biophysics, food science and technology, and electrical engineering.

Biomimetic
Sensor Technology

KIYOSHI TOKO

CAMBRIDGE
UNIVERSITY PRESS

CAMBRIDGE UNIVERSITY PRESS
Cambridge, New York, Melbourne, Madrid, Cape Town, Singapore, São Paulo

Cambridge University Press
The Edinburgh Building, Cambridge CB2 2RU, UK

Published in the United States of America by Cambridge University Press, New York

www.cambridge.org
Information on this title: www.cambridge.org/9780521593427

© Cambridge University Press 2000

First published 2000
This digitally printed first paperback version 2005

A catalogue record for this publication is available from the British Library

Library of Congress Cataloguing in Publication data

Toko, Kiyoshi, 1953–
 Biomimetic sensor technology / Kiyoshi Toko.
  p.   cm.
 Includes bibliographical references and index.
  ISBN 0 521 59342 5 (hardbound)
 1. Biosensors.   2. Chemoreceptors.   I. Title
   R857.B54T65   2000
   660.6′3 – dc21   99-25754   CIP

ISBN-13 978-0-521-59342-7 hardback
ISBN-10 0-521-59342-5 hardback

ISBN-13 978-0-521-01768-8 paperback
ISBN-10 0-521-01768-8 paperback

CONTENTS

v

PREFACE

Humans have five senses: sight, hearing, touch, smell and taste. The sensor plays the role of reproducing the five senses or surpassing them. In the history of sensor development, the sensors corresponding to the receptor parts of sight, hearing and touch have been developed for many years. By comparison, the sensors for simulating the senses of smell and taste have been proposed only recently in spite of the great demand for these sensors in the food industry and in environmental protection. The sensors for the senses of sight, hearing and touch respond to such single physical quantities as light, sound waves and pressure (or temperature), respectively. The end target in developing sensors for these parameters may be high sensitivity or selectivity for the physical quantity concerned, and this can be achieved by such means as semiconductor technology. On the contrary, many kinds of chemical substance must be assessed at once for smell and taste to be transformed into meaningful quantities to describe these senses. It has not been clear what materials can be adequately used to receive chemical substances that produce smell and taste.

This book deals with the development of what are known as the electronic nose and the electronic tongue, that is to say odor and taste sensors. These sensors have been developed on the basis of mechanisms found in biological systems, such as parallel processing of multidimensional information or by the use of biomaterials.

The first chapter is an introductory section dealing with the principles of measurements and multivariate analysis. The second chapter deals with reception mechanisms, which have been clarified so far in gustatory and olfactory systems. Biomimetic membrane devices to show excitation or behave as a switching element are reviewed in the third chapter. They have two key characteristics: self-organization, which appears at the equilibrium state and also far from equilibrium, and nonlinearity, which occurs at various levels. Biosensors, including the recent development of integrated microelectromechanical systems (iMEMS) are reviewed briefly in Chapter 4. The following chapters, 5–7, form the main part of this book and describe a novel strategy to express the subjective human sense on an objective, quantitative scale using several types of electronic nose and electronic tongue. The taste-sensing system developed by our research group utilizes lipid membranes, and many enthusiastic users have shown that it can really measure the taste of food-stuffs. These sensors can visualize taste and smell quantitatively and, hence, will make it possible to automate quality-control procedures for foods or cosmetics and facilitate amenity control of the environment. The last chapter is devoted to a description of the fusion of sensor technologies and the possibility of mimicing the reception mechanisms of the gustatory system, based on the results of the taste sensor.

The forthcoming multimedia world of the 21st century will have much greater demand for smart sensors and systems. I believe that this book presents a new stage in sensor technologies, aiming at intelligent sensors that can contribute to building a peaceful world.

I would like to express my gratitude to Dr S. Iiyama, Kinki University in Kyushu, Dr K. Hayashi, Kyushu University and Dr H. Ikezaki and Dr A. Taniguchi, Anritsu Corp. for their friendly cooperation for a long time. I am indebted to Prof. K. Yamafuji, who has encouraged me always. I wish to thank my colleagues in developing or using the taste-sensing system for their earnest, helpful efforts. My thanks for the typesetting are due to Ms S. Yanagisawa and Ms K. Asahiro. I would like to give sincere thanks to the publisher and the editorial/production staff, Dr S Capelin and Dr J. Ward.

Research into taste sensors for many years and the writing of this book would not have been possible without the warm support of my wife Kayano, to whom I wish to express my endless gratitude.

Kiyoshi Toko

1

Sensor and measurement

———

1.1 What is a sensor?

Humans have five senses: sight, hearing, touch, smell and taste, as illustrated in Fig. 1.1. These senses are very important because humans act after receiving information from the outside world.

Automatic doors open by checking for the presence of people using the device that corresponds to the sense of sight. These devices are common throughout the world. In fact, when we come across a nonautomatic door, we often assume that it must be self-operating and wait for a while for it to open, and then we become aware that it is nonautomatic. The infra-red light is usually used in detecting a person at automatic doors. The history of the automatic door is very old; in fact, it was used in Alexandria, the capital of ancient Egypt, more than 2000 years ago. It was made in such a way that it could open when fire was started at the altar in front of the door of the temple.

The sensor plays the role of reproducing the five senses or surpassing them. Figure 1.2 shows the correspondence between the biological system and the artificial system in the process of reception and the following action. The sensor is the device that mechanizes the ability of five organs, i.e., eye, ear, skin, nose and tongue, in the senses of sight, hearing, touch, smell and taste, respectively. With the development of computers, we often use the term sensor in the global sense by combining the data-processing part with the receptor part (i.e., the sensor in the narrow sense). In this case, the sensor plays roles of recognition as well as reception. This is the direction in which development of intelligent sensors is moving.

The actuator implies the mechanical part that moves the object by transforming the output from the sensor to rotation and displacement.

In the above example of the automatic door in ancient Egypt, the air under the altar is expanded by fire (sensor part); this causes water to flow into another vessel through a tube. The resulting water pressure moves the rotating apparatus of the door (actuator part).

1

Figure 1.1. Five senses of humans.

Table 1.1 summarizes the correspondence between the sensor and the five senses. "Odor sensor" and "taste sensor" are entered in the lists of the senses of smell and taste, respectively. This is based on the expectation that these two kinds of sense can be realized at the reception level if good sensing materials are developed successfully. As explained in Chapter 2, the primitive

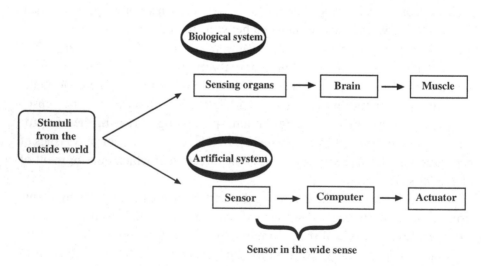

Figure 1.2. Correspondence between the biological system and the artificial system in the process of reception and the following action.

Table 1.1. *Five senses and sensors*

Five senses	Sensing organ	Object	Sensor	Principle
Sight	Eye	Light	Optical sensor	Photovoltaic effect (photon → electric change[a])
Hearing	Ear	Sound wave	Pressure sensor	Piezoelectric effect (sound wave → electric change)
Touch	Skin	Pressure	Pressure sensor	Piezoelectric effect (pressure → electric change)
		Temperature	Temperature sensor	Seebeck effect (temperature → electric change)
Smell	Nose	Chemical substances	Gas sensor	Chemical reaction (gas adsorption → electric resistance change)
			Odor sensor	Adsorption effect (mass change → frequency change)
Taste	Tongue	Chemical substances	Ion sensor	Selective ion permeation (ion → electric change)
			Taste sensor	Electrochemical effect (interactions → electric change)

[a] The term "electric change" implies changes in electric resistance, electric voltage or electric current.

discriminations of quality of taste and chemical substances to produce smell are made in gustatory and olfactory cells, respectively.

The sensor can surpass the ability of the five senses of humans in the following two ways. The first is that living organisms including human beings cannnot live in severe conditions; for example, humans cannot experience the temperature in a blast furnace themselves. It is the sensor that can achieve this. The second concerns the limited sensory range, depending on the species, of living organisms. For example, humans cannnot recognize propane gas or carbon dioxide and cannot hear the supersonic sound generated by a bat or a dolphin. Humans can perceive these quantities by using a sensor that surpasses their abilities.

In the history of sensor development, the sensors (in the narrow sense) corresponding to the receptor parts of sight, hearing and touch have been developed for many years. By comparison, the sensors for simulating the senses of smell and taste have been proposed only recently. This difference is a consequence of the mechanisms involved in these two kinds of sensor. In the senses of sight, hearing and touch, only one physical quantity – light, sound wave and pressure (or temperature), respectively – is received. Hence the sensors have only to transform this physical quantity to another tractable quantity such as an electric signal. On the contrary, many kinds of chemical substance must be assessed at once for smell and taste to be transformed into meaningful quantities to represent these senses.

As a result, sensors for sight, hearing and touch use popular materials such as semiconductors to receive the physical quantity. For smell and taste, however, it has been less clear what materials can be adequately used to receive the many kinds of chemical substance. For example, when we eat something, we express its taste by using the terms "sweet", "bitter" and "sour". It is said that there are about 1000 kinds of chemical substance in tea or coffee. We have no idea of the mechanism by which information contained in these substances is transformed into meaningful, simple expressions such as "sweet" and "bitter". However, it should be noted that the classification into these taste expressions is made at the first stage of reception of chemical substances by taste cells.

The sensors that play the roles of receptor in the senses of sight, hearing and touch are called physical sensors, because physical quantities are received. The sensors playing the role of receptor in the senses of smell and taste can be classified as the so-called chemical sensors. Nevertheless, the approach used to construct chemical sensors seems to have been almost the same as that in physical sensors, because it is based on high selectivity and sensitivity. One of the fruitful results is an enzyme sensor or an ion-selective electrode. They are very powerful and useful for detecting a specific chemical substance with high selectivity and sensitivity.

However, taste or smell cannot be measured if we fabricate many chemical sensors with high selectivity for different chemical substances when there are more than 1000 in one kind of foodstuff. The original role of smell and taste was to detect and assess information within a large mass of external information (enormous numbers of chemicals). The sense of smell is powerful in detecting the smell of an enemy or prey in the dark of a forest. The sense of taste is used to judge whether anything to be taken into the mouth is beneficial or poisonous for the body. There are too many kinds of chemical substance involved in producing taste and smell, and hence it seems important to obtain useful information quickly rather than to discriminate a single chemical species from others. This tendency is evident in unicellular living organisms, which have no sense of sight.

The above argument may need one comment concerning the sense of smell. Recent studies have revealed that there are numerous types of specific protein for odor molecules, as detailed in Chapter 2. In addition, there is no basic smell, whereas five basic taste qualities are known in the sense of taste. These facts may imply a large difference between the senses of smell and taste: the sense of smell has a specific property in real biological systems while the sense of taste keeps a nonspecific property to some extent. For this reason, the above argument concerning the fabrication of sensors may be good in the technical meaning for odor sensors, while for taste sensors it holds also in the biomimetic meaning.

As a summary, it can be said that the sensors that are used for smell and taste must have a fabrication principle that differs from the physical sensors or the conventional chemical sensors. Recently developed taste and odor sensors have outputs well correlated with the human sensory evaluations, and the taste sensor particularly has an intelligent ability to break down the information included in chemical substances to the basic information of taste quality. We are now standing at the beginning of a new age when sensors can be used to reproduce all five senses.

We have huge numbers of sensors that can be considered in terms of three factors: the materials selected for measuring the object, the purpose and use of the sensor and the physical effect or chemical reaction used in achieving the measurement (Fig. 1.3). For example, let us consider a situation where we measure pressure or acceleration. For this purpose, it seems good to use a

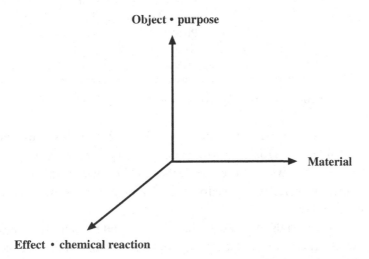

Figure 1.3. Various kinds of sensor with different effects and materials according to the object.

Table 1.2. *Various effects utilized in sensors*

Classification	Effect
Light	Zeeman
	Stark
	Doppler
	Raman
	Brillouin
	Nonlinear
	Optical parametric
Sound wave	Doppler
	Acoustoelectric
	Acoustomagnetic
	Masking
	Diffraction
Semiconductor	Tunnel
	Zener
	Electric field
	Gun
	Josephson
Magnetism	Superconducting quantum
	Barkhausen
Thermal	Seebeck
	Peltier
	Thomson
Photovoltaic	Photovoltaic
	Photoconductive
	Photoelectron emission
	Photoelectromagnetic
	Pockels
	Kerr
Piezoelectricity	Piezoelectric

From Ohmori (1994) with permission.

combination of the piezoelectric effect in Table 1.2 and the semiconductor material in Table 1.3. However, other combinations may be possible. To measure temperature, we can use the Seebeck effect, which transforms temperature change into electromotive force, and a metal that shows this effect as the sensing material.

What are the materials and effects that are adequate to measure the senses of taste or smell? It is not too much to say that a novel material and effect have been found in the development of sensor technology. If we can visualize taste and smell quantitatively, a new world will be opened.

Table 1.3. *Materials used for sensors*

Materials	Examples
Ceramics	SnO_2, ZnO, TiO_2, Al_2O_3, $BaTiO_3$, SiO_2
Semiconductor	Si, Ge, GaAs, InSb, CdSe
Metal	Cu, Fe, Ni, Pt
Polymer, enzyme[a]	Polyvinyl chloride, poly(pyrrole), glucose oxidase, valinomycin, crown ether
Living-organ material	Microbe, lipid, collagen, protein, catfish barb, frog skin, bacteriorhodopsin

[a] There is not a strict division between polymer/enzymes and material from living organisms.

1.2 Fundamental units

We can find the sentence "Time travels in diverse paces with diverse persons" in *As You Like It* written by Shakespeare. It implies a subjective factor in the concept of time, i.e., mental time. However, we can also have physical and objective time. We are always troubled by the dilemma between mental and objective time.

Measurement implies the ability to compare some obtained quantity with a fundamental value ("measure") deduced from some scale and to express it using the quantitative value. Let us consider a situation where we measure some quantity and get the value L, then the measured value becomes L/U if the fundamental value is U. If the quantity is concerned with length with a unit of m, we can say that the length of the object is L/U m.

While many units of length have been proposed so far, the most convenient way may be to use the human body (Fig. 1.4). Protagoras in ancient Greece said that "man is the measure of all things; of what is, that it is; of what is not, that it is not". While he meant the standard of the truth of all things by "measure of all things", we can apply this saying to our body. The fact that humans have tried to develop a common unit based on their bodies can also be found in a famous sentence in *Rongo* written by Koshi in ancient China; Koshi says that we know one *sun* by stretching our fingers, one *shaku* by our hand, and one *hiro* by our elbow. One *sun* is 3.03 cm, one *shaku* 30.3 cm, and one *hiro* 1.515 m. The unit cubit has an old history from over 4000 years ago in Mesopotamia and Egypt, and one cubit corresponds to 45 to 70 cm depending on the race and period. One yard, which might originate from double cubits, is equal to 3 feet (0.914 m). The same situation holds for the units inch, ell, fathom and so on.

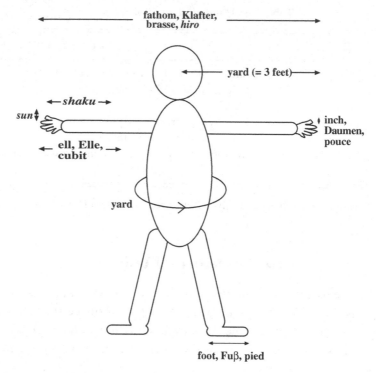

Figure 1.4. Scales based on the human body. The scales based on anatomy such as the length of elbow do not always have the same value, presumably because of differences in race, period and tradition.

With time, the unit was determined using a phenomenon that occurred repeatedly with an accurate period. The annual overflow of the Nile in ancient Egypt occurred when Sirius rises in the sky in the same direction as sunrise, and it promised a good harvest by carrying fertile soil from the upper stream. The unit "one year" (i.e., solar year) was determined from a period of the revolution of the earth.

In 1889, 30 meter prototypes (Fig. 1.5) were produced. One of them was assigned as an international meter prototype, the length of which at 0 °C was defined as 1 meter. The system unified on the base of this meter unit is the international system of metric units (SI: Système International d'Unités). The SI unit was accepted during a number of meetings of the Conférence Générale des Poids et Mesures.[2]

The SI system consists of seven base quantities (shown in Table 1.4), two additional quantities and numerous derived quantities. The units of the length and time have been improved and now they are defined within the framework of quantum mechanics. However, mass is based on the international kilogram

Table 1.4. *The seven base quantities of the SI system*

Base quantity	Name	Symbol	Definition
Length	meter	m	Distance travelled by light in vacuum during a time interval of $1/299\,792\,458$ second
Mass	kilogram	kg	Weight of an international platinum–iridium alloy prototype kept at the International Bureau for Weights and Measures in Sèvres, France
Time	second	s	Duration of $9\,192\,631\,770$ periods of electromagnetic radiation, corresponding to the transition between the two hyperfine levels of the ground state of the cesium-133 atom
Electric current	ampere	A	Current that produces a force of 2×10^{-7} newton per meter of length between two straight parallel conductors of infinite length placed 1 meter apart in a vacuum
Thermodynamic temperature	kelvin	K	$1/273.16$ of the thermodynamic temperature of the triple point of water
Amount of substance	mole	mol	Amount of a substance of a system containing as many elementary entities as there are atoms in 0.012 kilogram carbon-12
Luminous intensity	candela	cd	Intensity in a given direction of a source emitting monochromatic radiation of a frequency 540×10^{12} hertz with a radiant intensity in that direction of $1/683$ watt per steradian

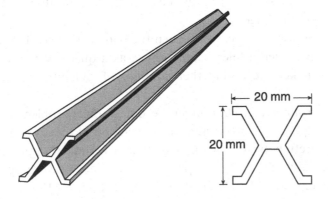

Figure 1.5. The meter prototype.

prototype, which is made of 90% Pt and 10% Ir, from 1889. It is cylindrical and is kept at the International Bureau for Weights and Measures (BIPM) in Sèvres, France.

Galileo found that the period of a simple pendulum is determined by the length of the pendulum and does not depend on the weight. The period of a 1 m simple pendulum is about 2.01 s, and hence half a period is about 1 s. It is easy to visualize the relationship between 1 m and 1 s.

The unit for the amount of substance, as atoms, molecules or ions, is the mole. Because there is an unimaginably large number of molecules, we use a "unit" called the Avogadro constant (originally termed Avogadro number), which is 6.02×10^{23}. The reference "prototype" is the number of atoms in 0.012 kg carbon-12. How large this number is can be grasped by the following hypothetical experiment. Let us fill a cup with water when we go to the sea. Then we put the water back into the sea and let the giant in the Aladdin's lamp stir the sea with a large spoon. After a sufficient stir, we again fill the cup with water from the sea. How many molecules that first filled the cup come back into the cup? The result is very surprising; some molecules will come back to the cup. As this example illustrates, the world we deal with is composed of an unimaginably large number of molecules.

Using a mol, we can define the unit for concentration in solution. A 1 M aqueous solution is 1 mol of molecules dissolved in 1 l (1000 cm^3) water. The NaCl content of sea water is 0.5 M (500 mM). The threshold for detection of chemical substances to produce taste and smell for ordinary individuals lies between 1×10^{-6} M (1 μM) and 0.1 M (100 mM). The decimal prefixes that are used to describe these multiples and submultiples of a thousand are given in Table 1.5.[2]

Figure 1.6 shows large numbers using hieroglyphics: 10^4 implies the form of bending a figure, 10^5 the form of some animal, a burbot, and 10^6 shows a person surprised. The sign for 10^7 may be an indication of unknown because the number is just too huge.

Figure 1.7 shows what we can see when the scale is increased. We human beings can imagine such a microscopic object as a quark, which cannot be observed by the naked eye, and further can think of such a huge object as the universe.

Recently, environmental pollution has become a social problem, and pollution of the atmosphere and rivers and the destruction of the ozone layer must be considered carefully from now on. We often hear the term ppm (parts per million) in our daily life. The unit ppm implies a ratio. For example, if 1 μg materials are included in 1 g water, it amounts to 1 ppm. More recently, we hear ppb (parts per billion). The appearance of this small unit may not be

Table 1.5. *Decimal prefixes*

Factor	Prefix	Abbreviation	Factor	Prefix	Abbreviation
10^{18}	exa	E	10^{-1}	deci	d
10^{15}	penta	P	10^{-2}	centi	c
10^{12}	tera	T	10^{-3}	milli	m
10^{9}	giga	G	10^{-6}	micro	μ
10^{6}	mega	M	10^{-9}	nano	n
10^{3}	kilo	k	10^{-12}	pico	p
10^{2}	hecto	h	10^{-15}	femto	f
10^{1}	deca	da	10^{-18}	atto	a

Figure 1.6. Hieroglyphics to express large numbers.

independent of the fact that humans produce chemical substances that are harmful in quantities as low as 10^{-9} (1 ppb).

1.3 Classification of measurement methods

Measurement methods can be divided into the following categories depending on the object and the properties of the measured quantity.

1.3.1 Direct and indirect measurements

Direct measurement implies a method to measure by directly comparing the measured quantity with the scale, for example reading the divisions of a scale in the case of measuring length. On the contrary, indirect measurement needs some calculation to get the result, for example to get velocity by the division procedure using the distance and the necessary time, which must be measured separately.

The accuracy of the indirect measurement is determined from the accuracy of each measurement. Let $x_1, x_2, \ldots, x_n$ denote the quantity obtained by the measurement and y denote the quantity to be obtained as a result, then y can be expressed by:[3,4]

$$y = f(x_1, x_2, \ldots, x_n), \tag{1.1}$$

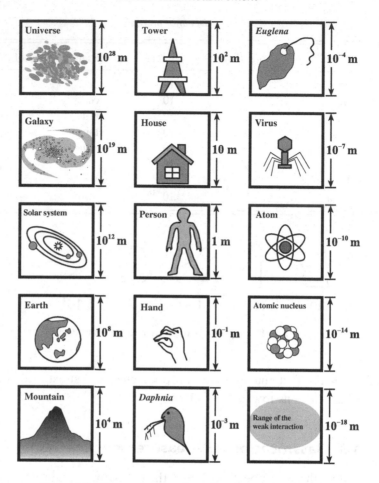

Figure 1.7. From microscopic scale to macroscopic scale.

where f implies some calculation procedure. The errors $\delta x_1, \delta x_2, \ldots, \delta x_n$ are generated in the measurements of $x_1, x_2, \ldots, x_n$, respectively. In this case, the resulting error of y becomes

$$\delta y = \frac{\partial f}{\partial x_1}\delta x_1 + \frac{\partial f}{\partial x_2}\delta x_2 + \cdots + \frac{\partial f}{\partial x_n}\delta x_n. \tag{1.2}$$

Each error δx_i affects the total error δy by the weight of $\partial f/\partial x_i$.

Let us consider one example of obtaining the moment of inertia of the cylinder with mass M, diameter d and length l. The moment of inertia y is given by

$$y = M\left(\frac{l^2}{12} + \frac{d^2}{16}\right). \tag{1.3}$$

From eq. (1.1) we get

$$\frac{\delta y}{y} = \frac{\delta M}{M} + \frac{8}{4 + 3d^2/l^2}\frac{\delta l}{l} + \frac{6}{3 + 4l^2/d^2}\frac{\delta d}{d}. \tag{1.4}$$

For the cylinder with $l = 10$ cm and $d = 1$ cm, $\delta y/y$ becomes

$$\frac{\delta y}{y} \simeq \frac{\delta M}{M} + 2\frac{\delta l}{l} + 0.015\frac{\delta d}{d}. \tag{1.5}$$

This result means that the measurement of the diameter d scarcely affects the moment of inertia.

As illustrated in this example, there are instances where we must make careful measurements and others where we can make somewhat rough measurements according to the particular factor in the indirect measurement.

1.3.2 Absolute and relative measurements

In absolute measurement, the measured quantity is an absolute value, as found when measuring light velocity ($c = \lambda f$) using the values of the wavelength λ and the frequency f. Here we can get the magnitude of velocity in itself if the wavelength and frequency are measured accurately. The same situation holds for the measurement of length.

By comparison, relative measurement implies that the method measures the difference from the value at a reference point, which can be set arbitrarily. A typical example is the measurement of electric potential at some point, because it depends on the reference point of zero electric potential.

Whereas the value at a reference point should be constant during the measurement, it changes in some cases. The difference method for cancelling the changeable value at the reference point is effective in this instance (see an example in Fig. 4.11).

1.3.3 Deflection and null methods

In the spring balance, we read the deflection of the sign attached to the spring, which is expanded by the weight of an object, as shown in Fig. 1.8(a). In the ammeter to measure electric current, we read the angle achieved by the rotational force caused by the electric current that induces the electromagnetic force in the magnet. These are deflection methods. Here we must obtain beforehand a relationship between the deflection and the measured quantities, i.e., a calibration line.

The null method is one whereby the unknown quantity is measured by balancing it with a known reference quantity. The indication is zero when

(*a*)

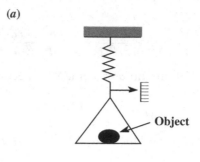

(*b*)

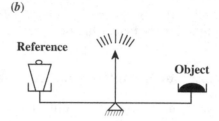

Figure 1.8. (*a*) The deflection method and (*b*) the null method.

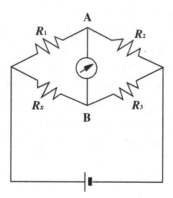

Figure 1.9. Wheatstone bridge.

the reference quantity is adjusted to the same value as the unknown quantity, as shown in Fig. 1.8(*b*).

One of the null methods is the measurement of electric resistance using the Wheatstone bridge, shown in Fig. 1.9. The unknown resistance R_x can be expressed using the known resistances R_1, R_2, R_3 by

$$R_x = \frac{R_1 R_3}{R_2}.$$
(1.6)

When the above equation holds, the electric current does not flow between points A and B because of zero electric potential difference.

In the null methods, a type of feedback procedure to adjust the known quantity is necessary.

1.4 Multiple regression analysis

In multiple regression analysis, we consider deriving a linear equation to give a relationship between several measured quantities. As one of the simplest examples, Fig. 1.10 shows the change in electric resistance $R(\Omega)$ of some material with increasing temperature $T(°C)$. Let us try to express the relationship between R and T by the following linear equation:

$$R = b_1 T + b_0, \tag{1.7}$$

where b_1 and b_0 are the constants to be determined to give a best fit to all the measured data. Since the number of pieces of data is generally greater than two, there cannot exist two constants (b_1, b_0) that will satisfy eq. (1.7) for all the data. So, what we can do is to search for the method for determining b_1 and b_0 in an approximate equation to reproduce all the data in the best way.

This subject can be extended to the case with more variables, $y, x_1, x_2, \ldots, x_p$. The expected value Y for the experimental value y can be expressed by the following equation:

$$Y = b_1 x_1 + b_2 x_2 + \cdots + b_p x_p + b_0. \tag{1.8}$$

This equation is called a multiregression equation with the explanatory variables x_i (predictor variables) and one dependent variable Y.

The two constants b_1 and b_0 in eq. (1.7) can be determined as follows. The error of each measurement i is given by

$$z_i = R_i - (b_1 T_i + b_0) \qquad \text{for} \qquad i = 1, 2, \ldots, n, \tag{1.9}$$

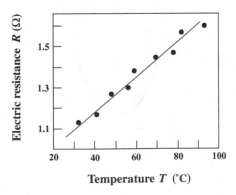

Figure 1.10. Change of electric resistance with temperature.

where R_i is the measurement value of the electric resistance and n is the total number of measurements. From the method of least squares, the error is smallest when $\sum z_i^2$ becomes a minimum with respect to b_0 and b_1 (see Fig. 1.11).

By minimizing $\sum z_i^2$ with respect to b_0 and b_1, we get

$$\frac{\partial}{\partial b_0} \sum_i z_i^2 = \sum_i [R_i - (b_1 T_i + b_0)] = 0,$$

$$\frac{\partial}{\partial b_1} \sum_i z_i^2 = \sum_i [R_i - (b_1 T_i + b_0)] T_i = 0. \tag{1.10}$$

From this equation, we get

$$b_0 = \frac{\sum_i R_i \sum_i T_i^2 - \sum_i R_i T_i \sum_i T_i}{n \sum_i T_i^2 - (\sum_i T_i)^2},$$

$$b_1 = \frac{n \sum_i R_i T_i - \sum_i R_i \sum_i T_i}{n \sum_i T_i^2 - (\sum_i T_i)^2}. \tag{1.11}$$

The above case comprises only one explanatory variable. In more general cases with many predictor variables, the following set of equations holds for each measurement:

$$
\begin{aligned}
Y_1 &= b_1 x_{11} + b_2 x_{21} + \cdots + b_p x_{p1} + b_0, \\
Y_2 &= b_1 x_{12} + b_2 x_{22} + \cdots + b_p x_{p2} + b_0, \\
&\vdots \\
Y_n &= b_1 x_{1n} + b_2 x_{2n} + \cdots + b_p x_{pn} + b_0.
\end{aligned} \tag{1.12}
$$

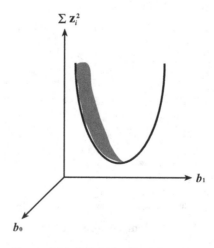

Figure 1.11. Minimization of errors.

Minimization of $\sum z_i^2$ leads to

$$\sum_i (y_i - Y_i) = 0, \tag{1.13a}$$

$$\sum_i (y_i - Y_i)x_{ji} = 0. \tag{1.13b}$$

Equation (1.13a) implies that the average value $\bar{y}(= \sum y_i/n)$ of the measured quantity is equal to the average of the expected value $\bar{Y}(= \sum Y_i/n)$. By multiplying b_j in eq. (1.13b) and making a summation about j, we get

$$\sum_i (y_i - Y_i)Y_i = 0, \tag{1.14}$$

by taking into account eq. (1.13a). Using eq. (1.14) we get

$$\sum_i (y_i - \bar{y})^2 = \sum_i (Y_i - \bar{Y})^2 + \sum_i (y_i - Y_i)^2. \tag{1.15}$$

This equation implies that the square sum (S_T) of the observed value from which the average is subtracted is equal to the sum (S_R) of the square sum of expected value, from which the average is subtracted, and that of error (S_E).

$$\frac{S_E}{S_T} = 1 - \frac{S_R}{S_T}. \tag{1.16}$$

For the smaller S_E, the multiregression equation can explain the experimental data.

The ratio S_R/S_T can be expressed by

$$\frac{S_R}{S_T} = r^2, \tag{1.17}$$

where r is the multicorrelation coefficient defined by

$$r = \frac{\sum_i (y_i - \bar{y})(Y_i - \bar{Y})}{\sqrt{\sum_i (y_i - \bar{y})^2 \sum_i (Y_i - \bar{Y})^2}}. \tag{1.18}$$

From the above equations, it can be concluded that if r is near 1, the multiregression equation can also explain the experimental data.

The multiregression analysis is used with the smallest number of predictor variables. In general cases, n is larger than p in eq. (1.12). A peculiar situation arises in the case of $n = p + 1$. The number of unknown coefficients $b_0, b_1, \ldots, b_p$ agrees with the number of equations. So, b_i can be solved exactly except for a special case where one equation coincides with another equation because of the same data values that were obtained casually. This leads to zero errors; that is, eq. (1.8) can express the observed data strictly. Therefore, it is expected that we can obtain a multiregression equation to explain the observed data only if we increase the number of explanatory variables that have no relationship with the phenomenon concerned. Then, r^2 increases artificially by adding meaningless explanatory variables. This is very strange. It is an unfavorable artifact brought about by the increasing degree of space constructed from the explanatory variables, i.e, the degree of freedom.

   To prevent the meaningless increase in correlation coefficient, the adjusted r^2 is proposed as follows:

$$\hat{r}^2 = 1 - \frac{n-1}{n-p-1}(1 - r^2). \tag{1.19}$$

As can be understood from the above equation, $\hat{r}^2$ can decrease if meaningless explanatory variables are added to the multiregression model. Using $\hat{r}^2$, we can compare the correlation between cases with different numbers of explanatory variables.

   In some cases, cross-terms consisting of x_i and x_j are necessary to express Y. The same procedure is possible by regarding the cross-term as a new explanatory variable.

1.5 Principal component analysis

When we make measurements, we usually wish to express some components quantitatively. In the case of direct measurement, the measured quantity is the object itself. In the case of indirect measurement, however, the object (dependent variable) is expressed by a set of measured quantities (explanatory variables). The method to achieve this was described in Section 1.4. In this section, we treat the case where there is no dependent variable, which occurs frequently. For example, let us assume that we have a detailed investigation in a hospital. The investigation produces a number of measurements, total bilirubin, GOT, GPT, . . . , in the list of functional tests (the explanation of GOT, GPT is unnecessary; we only need to know that they are important quantities measured in assessing our health). When we want to know the hepatic function, we must search the equation by which the hepatic function is expressed using the above measured quantities.

   Let us consider an example, shown in Table 1.6, where scores of language and mathematics of 16 students are listed. Using this result, it may be possible to say whether student E is good at language or mathematics. It implies the formulation to express the term "good at language" (or "good at mathematics") using the score of language x_1 and the score of mathematics x_2. In this case, the decrease in information occurs always because the two variables (scores of language and mathematics) are reduced to one variable ("good at language"). Our intention is to adopt the method to obtain useful information

Table 1.6. Scores of language and mathematics of 16 students

Student	A	B	C	D	E	F	G	H	I	J	K	L	M	N	O	P
Language	78	75	61	88	86	65	63	87	75	67	80	83	87	90	72	82
Mathematics	71	73	79	67	65	79	82	74	76	78	72	70	64	65	76	80
PC1	1.5	−2.0	−17.2	12.2	11.4	−13.7	−16.9	8.0	−3.5	−11.5	2.8	6.4	12.8	15.0	−6.1	0.7

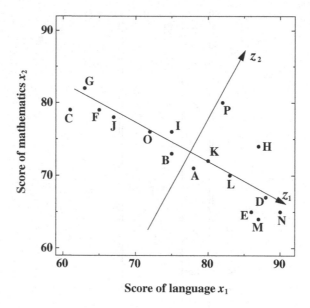

Figure 1.12. Scattering diagram of the scores of language and mathematics of 16 students.

without losing information.[5,6] This method is called a principal component analysis, which is often abbreviated as PCA.

The scores in Table 1.6 are rewritten in Fig. 1.12. By examining this figure, it would appear that the students who are good at mathematics are bad at language, while the students who are good at language are bad at mathematics. Therefore, it may be natural to define the lower right direction as the term "good at language" by drawing a straight line, as shown in Fig. 1.12. It means the introduction of a new variable z_1:

$$z_1 = a_{11}x_1 + a_{12}x_2, \tag{1.20}$$

where x_1 and x_2 are the values after the average was subtracted from the original values. Eq. (1.20) is nothing but rotation of axis.

We must do our best in determining a_{11} and a_{12} without losing information. It implies that the new axis is determined by the direction of the large spread of data points. To keep the magnitude in the transformation in eq. (1.20), the following condition is set:

$$a_{11}^2 + a_{12}^2 = 1. \tag{1.21}$$

Figure 1.13 shows schematically a relationship among the original information, the new information and loss of information. We get

$$\overline{OP_i}^2 - \overline{OQ_i}^2 = \overline{PQ_i}^2. \tag{1.22}$$

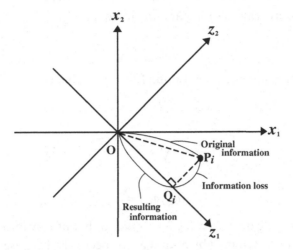

Figure 1.13. Concept of principal component analysis to obtain meaningful information.

This equation implies that $\overline{OQ}_i^2$ can be maximized in order to minimize $\overline{PQ}_i^2$ (loss of information). This is a problem in that we maximize $\overline{OQ}_i^2$ by summing up over 16 pieces of data under the condition of eq. (1.21).

For this purpose, the Lagrange's method of undetermined multipliers is powerful. We maximize the following equation including the undetermined multiplier λ:

$$g(a_{11}, a_{12}) = \sum_{i=1}^{N}(a_{11}x_{1i} + a_{12}x_{2i})^2 - \lambda(a_{11}^2 + a_{12}^2 - 1), \qquad (1.23)$$

where N is equal to 16 in the present example. By maximizing eq. (1.23) with respect to a_{11} and a_{12}, we get

$$\frac{\partial g}{\partial a_{11}} = s_{11}a_{11} + s_{12}a_{12} - \lambda a_{11} = 0,$$

$$\frac{\partial g}{\partial a_{12}} = s_{12}a_{11} + s_{22}a_{12} - \lambda a_{12} = 0, \qquad (1.24)$$

with

$$s_{11} = \sum_i x_{1i}^2,$$

$$s_{12} = \sum_i x_{1i}x_{2i}, \qquad (1.25)$$

$$s_{22} = \sum_i x_{2i}^2.$$

The quantities obtained by dividing s_{11} and s_{22} by N are called a variance and the quantity s_{12}/N is called a covariance.

For the non-zero values of a_{11} and a_{12} in eq. (1.24),

$$\begin{vmatrix} s_{11} - \lambda & s_{12} \\ s_{12} & s_{22} - \lambda \end{vmatrix} = 0 \tag{1.26}$$

is a necessary and sufficient condition. The eigenvalue is given by

$$\lambda_1 = \left[s_{11} + s_{22} + \sqrt{(s_{11} - s_{22})^2 + 4s_{12}^2} \right]/2,$$

$$\lambda_2 = \left[s_{11} + s_{22} - \sqrt{(s_{11} - s_{22})^2 + 4s_{12}^2} \right]/2. \tag{1.27}$$

Since the relation

$$g(a_{11}, a_{12}) = \lambda \tag{1.28}$$

holds, the magnitude of eigenvalue λ is equal to that of new information. The axes z_1 and z_2 are determined by λ_1 and λ_2, respectively, because $\lambda_1 > \lambda_2$. The following expressions can be obtained for a_{11} and a_{12}, if we take $a_{11} > 0$:

$$a_{11} = \frac{|s_{12}|}{\sqrt{(\lambda_1 - s_{11})^2 + s_{12}^2}},$$

$$a_{12} = \frac{\lambda_1 - s_{11}}{s_{12}} a_{11}. \tag{1.29}$$

For the example in Table 1.6, we get $a_{11} = 0.88$ and $a_{12} = -0.48$. The line z_1 in Fig. 1.12 is drawn using eq. (1.20) with these values. The positive value of a_{11} and negative value of a_{12} means z_1 increases for better at language and worse at mathematics. It really expresses the students who are better at language than at mathematics. The fact of $|a_{11}| > a_{12}$ implies that the axis z_1 is determined more by the score of language than that of mathematics. This axis has the largest information contained in x_1 and x_2, and hence it is called a first principal component (PC1). The axis z_2 determined from another eigenvalue λ_2 is called a second principal component (PC2).

As shown in eq. (1.28), the magnitude of eigenvalue reflects information. The ratio of the eigenvalue to the sum of all the eigenvalues is called a contribution rate. In the present example, the contribution rates are 92.9% and 7.1% for PC1 and PC2, respectively. Therefore, PC1 has almost all the information.

Figure 1.12 implies that the student E is very good at language because his PC1 value is high. On the contrary, the student G is very bad at language because of the low score of PC1. In this way, we could extract information regarding the term "good at language" from two scores of language and mathematics.

The axis z_2 is given by

$$z_2 = a_{21}x_1 + a_{22}x_2, \tag{1.30}$$

where the vector (a_{21}, a_{22}) belonging to the eivenvalue λ_2 becomes $(0.48, 0.88)$. The fact that $a_{21} > 0$ and $a_{22} > 0$ means that z_2 reflects the students who are relatively good at both language and mathematics. The student P takes the largest value of z_2. The value of PC2 can be regarded as a kind of total score of examinations.

This can be understood clearly from the next example, shown in Table 1.7. By using the same procedure as above, we get

$$A = \begin{pmatrix} 0.73 & 0.68 \\ -0.68 & 0.73 \end{pmatrix}, \tag{1.31}$$

where

$$\mathbf{z} = A\mathbf{x} \tag{1.32}$$

with $\mathbf{z} = (z_1, z_2)^t$ and $\mathbf{x} = (x_1, x_2)^t$, the superscript t implying the transpose vector. The contribution rates to PC1 and PC2 are 77.6% and 22.4%, respectively. As can be understood from the expression of eq. (1.31), z_1 becomes the sum of x_1 and x_2 with different weights.

This expression is similar to the simple sum of scores and sometimes is more effective for evaluating the result of examinations. If all the students get the same score, for example 80 for mathematics, the addition of the score in mathematics to those for examinations of other subjects is meaningless. In this case, the score in mathematics has no value in evaluating differences in the ability of students (no ability difference or problems too bad). If the PCA is made, one eigenvalue becomes zero because there is no dispersion of the score in mathematics. As a result, PC1 is determined by other examinations.

The PCA was made here using the variance and covariance. However, it is not adequate to perform the analysis when there are variables with different characteristics, for example weight (kg) for x_1 and height (m) for x_2 in assessing obesity. If the unit g is chosen for the weight, the dominant term becomes the weight, and obesity can be determined by only the weight x_1 independent of the height x_2. In such a case, the method using the nondimensional quantities is usually convenient: dividing x_i by the root of variance of x_i. It leads to $s_{ii} = 1$, and s_{ij} becomes the correlation coefficient r_{ij}.

Two variables were reduced to one variable in the above example. Of course, the same procedure can be applied to more general cases including many variables. The ultimate goal of sensor technology is to mimic the human senses and exceed them. For this aim, we sometimes get necessary information by using several sensors of different types and properties. In this case, we must relate all the information from these sensors to the intended quantities without losing information. The PCA is very useful for this purpose. In some cases, however, a relationship between input and output of the sensor

Table 1.7. *Another example of scores of language and mathematics*

Student	A	B	C	D	E	F	G	H	I	J	K	L	M	N	O	P
Language	80	72	75	87	63	75	66	85	78	67	68	82	80	77	71	85
Mathematics	64	66	80	85	62	73	76	76	79	70	62	70	62	68	75	83
PC1	-2.8	-7.2	4.6	18.8	-16.5	-0.2	-4.7	9.1	6.1	-8.1	-12.8	2.8	2.7	-2.2	-1.7	13.9

becomes strongly nonlinear. A neural network algorithm can be used in those cases, as will be found in Chapter 5 for odor sensors.

REFERENCES

1. Ohmori, T. (1994). *Basis and Application of Sensor Technology*. Saiwai Shobo, Tokyo [in Japanese].
2. Klaassen, K. B. (1996). *Electronic Measurement and Instrumentation*. Cambridge University Press, Cambridge.
3. Mashima, S. and Isobe, T. (1980). *Keisokuho Tsuron* (*General Methods of Measurements*). Tokyo University Press, Tokyo [in Japanese].
4. Toko, K. and Miyagi, K. (1995). *Sensa Kogaku* (*Sensor Engineering*). Baifukan, Tokyo [in Japanese].
5. Dillon, W. R. and Goldstein, M. (1984). *Multivariate Analysis*. Wiley, New York.
6. Weimar, U., Vaihinger, S., Schierbaum, K. D. and Göpel, W. (1991). In *Chemical Sensor Technology*, Vol. 3, ed. Seiyama, T. Kodansha, Tokyo, p. 51.

2

Chemical senses

———

2.1 Chemoreception

The senses of taste and smell are the chemical senses induced when chemical substances interact with the tongue and the nasal cavity, respectively. Physical quantities such as light (photon), sound wave and pressure (or temperature) are received in the senses of sight, hearing and touch, respectively, as shown in Table 1.1.

Chemoreception occurs even in unicellular organisms, although some living organisms such as deep-sea fish have no sense of sight among the physical senses. Protozoa such as amoebae and microbes such as colon bacilli show chemotaxis; they gather and escape from some chemical substances. The former is called positive chemotaxis and the latter negative chemotaxis. Colon bacilli show positive chemotaxis for amino acids tasting sweet and negative chemotaxis for chemical substances tasting strongly bitter or sour. This behavior is quite reasonable because substances tasting sweet become energy sources for living organisms whereas substances tasting strongly bitter or sour are often harmful.

As the above examples indicate, our likes and dislikes for chemical substances (i.e., foodstuffs) can be considered as an essential matter related to our safety. The senses of taste and smell are those used for checking the safety of substances that are ingested, and hence they have developed in higher animals in the same way as in unicellular living organisms that survive using chemical senses. The development of taste and odor sensors is a growing area of biomimetic technologies intended to mimic the original purpose of biological systems.

2.2 Biological membranes

The cell is enveloped by a biological membrane that is composed of proteins and the lipid bilayer. Figure 2.1 shows a fluid mosaic model proposed by

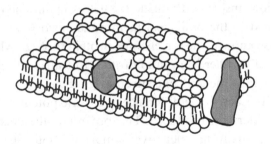

Figure 2.1. Fluid mosaic model of the biological membrane.

Singer and Nicolson in 1978,[1] where proteins can move freely in the "sea" of lipid bilayer.

A lipid molecule comprises a hydrophilic group that is soluble in water and a hydrophobic part that usually contains two hydrocarbon chains, which are insoluble in water. Because of this structure, lipid molecules can automatically form the lipid bilayer by gathering the hydrophobic chains inside and facing the hydrophilic group to the water phase. This is a kind of self-organization, which may play an important part in developing biomimetic devices.

As one example of a lipid, let us show the chemical structure of phosphatidylcholine, which is a main constituent of biological phospholipid.

$$CH_3CH_2CH_2 \cdots CH_2COO \diagdown \atop CH_3CH_2CH_2 \cdots CH_2COOCH_2 \diagup \ CHCH_2OPOCH_2CH_2N^+ - CH_3$$

The two long chains of carbon and hydrogen on the left are the hydrophobic hydrocarbon chains; the choline group made of phosphate and ammonium, which are charged negatively and positively near neutral pH, respectively, is hydrophilic. Such a lipid is sometimes symbolized by

Ions cannot move through the lipid bilayer, which can be regarded as almost an insulator because its electric resistance reaches over $1\,G\Omega/cm^2$. Potassium ions are mainly contained inside the cell, whereas Na^+ is dominant outside. As our blood contains a large amount of NaCl, this may support the suggestion that living organisms originated in an ancient sea. The unbalanced distribution of ions inside and outside the cell is maintained with the aid of proteins contained in the biological membrane. Two kinds of protein that transport ions are known; one is a channel protein through which a particular ion can move and another is a pump protein to transport ions using energy (usually ATP) against the ion concentration gradient.

As a result of this activity, a difference in electric potential between the cell interior and the exterior is produced. This potential difference, called a membrane potential, usually has negative values of about 80 mV because the membrane in the resting state is permeable to K^+: this is a stationary state realized without external stimuli. If a stimulus such as an electric current is applied to the membrane of a nerve fiber, the membrane potential changes drastically, as shown in Fig. 2.2. This transient change is first brought about by Na^+ influx and later by K^+ efflux and can be transmitted along the nerve fiber with a velocity of about 10 m/s, finally reaching the brain. In the resting state, the cell interior remains polarized in relation to the exterior; however, when a stimulus occurs, the changes are called depolarization, because the membrane potential gets near to zero. This transient depolarization is called an action potential or spike (or sometimes impulse). In this way, the nerve cell shows "excitation" by generating a series of action potentials in response to stimuli. Anesthetics suppress excitation, which results in paralysis of any part of the body treated with a local anesthetic such as tetracaine.

Living organs utilize the excitation in information transmission and processing. The number of spikes increases with the logarithm of stimulus intensity. Responses in the gustatory receptors of many animals increase linearly with

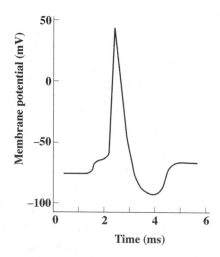

Figure 2.2. Generation of an action potential.

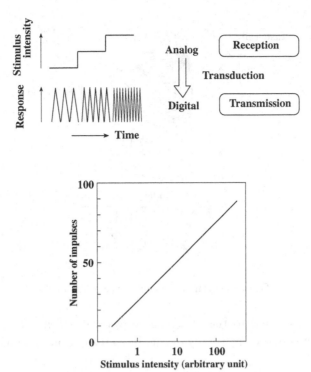

Figure 2.3. Transduction from the stimuli to the spike trains.

the concentration of taste substances. This empirical law is well known as Weber–Fechner's law (Fig. 2.3).

2.3 From reception of taste substances to perception in the brain

We have many sandy grains, called papillae, on our tongue. One papilla contains several to a few hundred taste buds depending on the kind of papilla. An adult has about 9000 taste buds 60–80 µm in length and at maximum 40 µm in diameter. The number of taste buds differs in different species, e.g., about 100 000 in the catfish, about 25 000 in the cow and about 7000 in a rabbit.

A taste bud is composed of several taste cells (gustatory cells), as shown in Fig. 2.4. Chemical substances are received at the biological membrane of the taste cell, which is not a nerve cell but rather an epidermal cell. The reception mechanism is not yet clear and will be discussed in more detail below. The membrane potential of a taste cell is changed at the first stage of chemical reception. Figure 2.5 shows reception of chemical substances by taste cells and olfactory cells. Although the detailed transduction mechanism is not clear, a

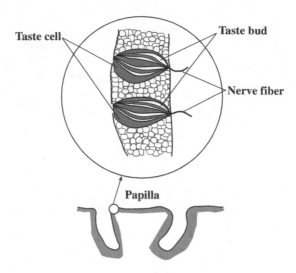

Figure 2.4. Papillae, taste buds and taste cells.

nerve fiber connected to the taste cell shows excitation. The connection is made using a synapse, as usually occurs in transduction of signals in animals (Fig. 2.6).

Figure 2.7 shows the number of impulses in 50 chorda tympani fibers of a rat.[2] We can see that one nerve fiber does not necessarily have the information of only one taste quality. The nerve fiber shows nonselective, nonspecific response to each taste quality so that one fiber, for example H in Fig. 2.7, may respond to NaCl, HCl and quinine. The pattern constructed from the

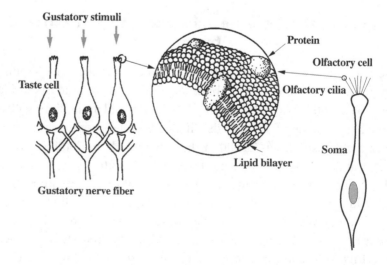

Figure 2.5. Reception of taste and smell.

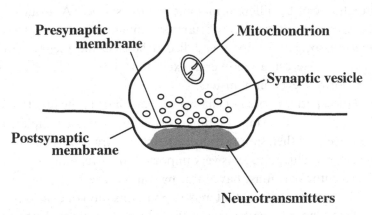

Figure 2.6. Information transduction at the synaptic junction.

number of impulses in 50 nerve fibers shows a quite different pattern for NaCl and sucrose. The patterns for HCl and quinine are also characteristic, although these two are somewhat similar compared with NaCl and sucrose. This result supports the across-fiber pattern theory proposed by Erickson *et al.* in 1965.[3] According to this theory, the taste quality is distinguished using the overall excitation pattern of nerve fibers.

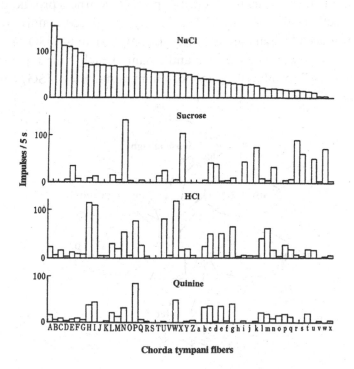

Figure 2.7. Response patterns constructed from 50 nerve fibers in a rat (Ogawa *et al.*, 1968).

This situation can be different in some animal species. A group of nerve fibers that respond to only sweet substances is found in hamsters. The theory based on these specialized fibers is called a labeled-line theory. However, transmission of taste using the overall excitation pattern can be considered to play a major role even in hamsters.

The excitation pattern does not change very much from the first to the third neurons. It implies that the information processing is almost completed at the first neuron level and then simply transmitted to the brain through the second and third neurons. This fact seems very important in developing a taste sensor in that the receptor part must have information of taste.

Figure 2.8 shows the factors that make up the sensation of deliciousness felt by us. We take dinner by gathering all the information from the five senses, i.e., the senses of taste, smell, touch, hearing and sight. Furthermore, deliciousness is affected by the mental and physical condition of the individual, and it strongly depends on the food culture of a country or race. Because of this, it seems extremely difficult to provide some measure of deliciousness of foods using artificial sensing systems. The taste sensor mentioned in Chapter 6 has been developed to measure the taste received at the tongue. It may contribute to objective quantification of deliciousness to some extent.

Five taste qualities are acknowledged at present: sourness produced by H^+ from HCl, acetic acid, citric acid, etc.; saltiness produced mainly by NaCl; bitterness produced by quinine, caffeine, L-tryptophan and $MgSO_4$; sweetness from sucrose, glucose, L-alanine, etc.; and umami, which is the Japanese term for "deliciousness", produced by monosodium glutamate (MSG) contained

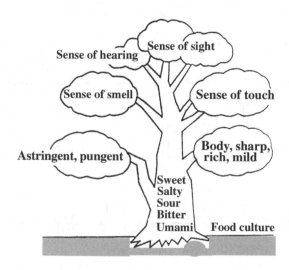

Figure 2.8. Factors of deliciousness.

Table 2.1. *Five basic tastes*

Taste	Chemical substances	Threshold	Physiologic significance
Sweetness	Sucrose, glucose, glycine, L-alanine, aspartame	10 mM (sucrose)	Energy source
Saltiness	NaCl, KCl, KBr	10 mM (NaCl)	Mineral supply
Sourness	H^+ produced by HCl, citric acid, acetic acid	0.9 mM (HCl)	Signal of rotting
Bitterness	Quinine, caffeine, picric acid, $MgSO_4$, L-tryptophan	8 μM (quinine)	Signal of poison
Umami	Monosodium glutamate (MSG), disodium inosinate (IMP), disodium guanylate (GMP)	2 mM (MSG)	Supply of amino acid and nucleotides

mainly in seaweeds such as tangle, disodium inosinate (IMP) in meat and fish and disodium guanylate (GMP) in mushrooms.[4–6] Umami is now acknowledged as a fifth basic taste.

Table 2.1 summarizes the chemical substances producing taste, the threshold detectable by humans and the physiologic significance. Here, the threshold means the minimum concentration at which a difference from water can be detected. By comparison, a recognition threshold is the concentration where the taste quality can be recognized. The recognition thresholds of sucrose, NaCl and quinine are 170 mM, 30 mM and 30 μM, respectively, which are higher than the thresholds listed in Table 2.1.

It is well known that sweet substances such as sucrose suppress bitterness produced by substances such as quinine, and coexistence of IMP or GMP enhances umami produced by MSG. The former is called a suppression effect, while the latter is a synergistic effect. Pursuing the mechanism of such taste interactions is one of the main themes of olfactory neurophysiologic and biochemical fields. Development of the taste sensor using lipid membranes may make a substantial contribution in this area.

Amino acids are important because they produce taste in many kinds of food, typically in fermented foods such as beer, wine, sake and soybean paste (miso).[6–8] Amino acids are used in processed foods because they enhance the nutritive value of many foods and also modify the natural taste characteristics of many foodstuffs, as is well known with MSG, which is a substance that shows the independent fifth taste, umami. The taste of an amino acid changes only if one group, R, is changed (Table 2.2).

$$R - - - \overset{\displaystyle NH_2}{\underset{\displaystyle H}{\overset{\big\blacktriangledown}{\underset{\big\blacktriangle}{C}}}} - - - - COOH$$

Glycine tastes sweet, L-valine sweet and bitter simultaneously, and L-trypto-phan tastes bitter. If the taste of amino acids contained in foodstuffs are measured, we can discuss the taste quantitatively and furthermore monitor processes such as fermentation.

Figure 2.9 summarizes our knowledge on the gustatory reception at the biological membrane and the following transduction.[9-14] However, the mechanism may partly depend on the animal species. Sweet substances are

Table 2.2. *Taste of amino acids: the size of circle implies the relative magnitude of five basic taste qualities shown by each amino acid*

Amino acid	R	Salty	Sour	Sweet	Bitter	Umami
Glycine	H			◎		
L-Alanine	CH_3			◎		
Monosodium L-aspartate	$NaOOC-CH_2$	◎				◎
L-Histidine monohydrochloride	$CH{=}C-CH_2$, N, NH, CH	○	◎			
L-Methionine	$CH_3-S-CH_2-CH_2$			○	◎	○
L-Lysine monohydrochloride	$H_2N-(CH_2)_4$ HCl			○	◎	
L-Valine	CH_3-CH, CH_3			◎	◎	
L-Tryptophan					◎	

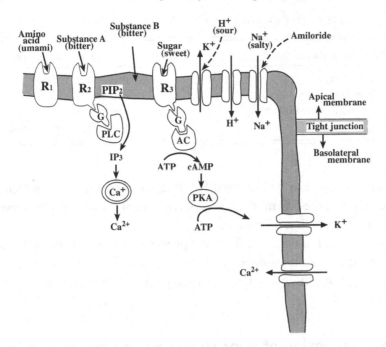

Figure 2.9. Model of gustatory reception and transduction. G, GTP-binding protein; AC, adenylate cyclase; PLC, phospholipase C; PKA, protein kinase A; IP$_3$, inositol trisphosphate; PIP$_2$, phosphatidylinositol 4,5-bisphosphate.

considered to be received by two different pathways. One is the cyclic AMP (cAMP)-mediated pathway for sugars, and the other is the inositol trisphosphate (IP$_3$)-mediated pathway for synthetic sweetners. Some bitter substances are received by a similar IP$_3$-mediated pathway, but another pathway without second messengers is also suggested. There is a possibility that bitter substances such as quinine are received at the hydrophobic part of the lipid bilayer and result in a membrane potential change. This hypothesis is supported from an experiment using liposomes[12] and also the results using a multichannel taste sensor with lipid/polymer membranes, as detailed in Chapter 6.

An amiloride-sensitive ion channel may act as the Na$^+$ receptor. Influx of Na$^+$ induces the depolarization. However, the transduction mechanism is unclear for other alkali cations, because large salts such as FeCl$_3$, which cannot move through the ion channel, induce a large response. Two possibilities are considered for the transduction of sourness. One is the direct influx of H$^+$ and the other is the suppressive action of H$^+$ on the K$^+$ channel, depending on the species. For these two kinds of taste quality, neither cAMP nor IP$_3$ concentration changes.

Table 2.3. *Comparison between taste and odor substances*

Taste substances	Odor substances
Received at the tongue	Detected by the nose
Nonvolatile in most cases	Volatile
Highly polar and soluble in water	Low polarity and soluble in oil
Exist in five basic tastes	Numerous types of quality

The umami taste can be imparted by glutamate, receptors for which have been identified in the central nervous system; receptors of a similar type may be involved in reception of umami.

The depolarization resulting from the processes following the taste stimuli causes opening of voltage-dependent Ca^{2+} channels, which leads to Ca^{2+} influx. Neurotransmitters are then released from synaptic vesicles to the gustatory nerve cell.

2.4 From reception of odor substances to perception in the brain

The sense of smell allows information of chemical environmental conditions to be assessed from a distance. Numerous types of odor molecule can be detected with extremely high sensitivity. Wandering salmon come back to the home river by remembering the smell of the river where they were born and bred.

In the field of foodstuffs, the term "flavor" is used to express comprehensively taste and smell.

It is said that there are about 500 000 types of chemical substance producing smell. Amoore in the 1960s classified them into seven classes according to their external form and electric charge:[15] camphoraceous, etheric, pepperminty, musky, floral, pungent and putrid. At present, however, many researchers believe that no basic smells exist. Nevertheless, Amoore's hypothesis may be valuable in the sense that it allows a hypothesis of receptor proteins specific for odor molecules, as mentioned below.

Many odor substances are low-molecular-weight compounds that are volatile and hydrophobic with no electric charge. Comparison between taste and odor substances is made in Table 2.3.

Figure 2.10 illustrates the process of reception of odorants by olfactory cells and then the transduction of information to the brain. As the olfactory cells are nerve cells, they directly elicit spike trains soon after the olfactory

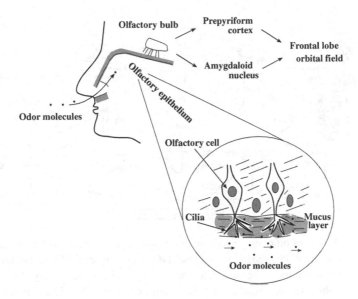

Figure 2.10. Reception of odor substances by the nose.

cilia receive odor molecules, which happens when these have dissolved in the mucus layer.

Tonoike and co-workers have summarized the steps in olfactory transduction.[16] This process occurs in four stages: (1) the generation of membrane depolarization (receptor potentials, known as an electro-olfactogram (EOG)) following the reception of odorant molecules by cilia; (2) the transmission of receptor potentials from the cilia to the soma; (3) the conversion of the EOG into spike trains at the axonal basal part of the cell body; and (4) membrane depolarization and recovery from the activated processes at cilia and the other parts.

Figure 2.11 shows a model of olfactory transduction based on present knowledge.[16] A powerful candidate for the receptors was discovered by Buck and Axel in 1991.[17] This is a multigene family encoding olfactory-specific proteins with seven transmembrane domains and acting through coupling to GTP-binding proteins (G-proteins). A specific odor molecule is thought to be trapped by a pocket structure surrounded by the seven transmembrane domains. It is now considered that there are about 1000 receptor proteins coded by different genes for different groups of odor molecules.

The importance of the hydrophobic part of the biological membrane (the lipid bilayer) is also indicated by experiments using liposomes.[18]

A cAMP-mediated pathway is now confirmed as one of the second messenger systems. The transduction is as follows: receptor protein → G-protein → adenylate cyclase (AC) → cAMP formation → cAMP-gated channel →

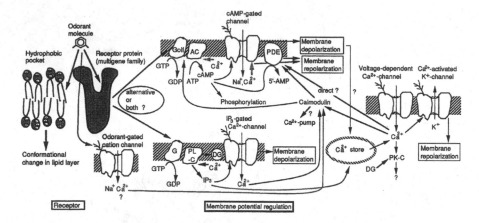

Figure 2.11. Model of olfactory reception and transduction.[16] Golf, olfactory-specific GTP-binding protein; PDE, phosphodiesterase; DG, diacylglycerol; PK-C, protein kinase C. Other abbreviations as in Figure 2.9. Striped arrows indicate inhibitory effects, while thick arrows show facilitative effects. Arrows marked with "?" indicate unidentified relationships.

membrane depolarization → cAMP hydrolysis. An IP_3-mediated transduction is also suggested from the identification of a GTP-dependent phosphatidylinositol 4,5-bisphosphate phosphodiesterase (PIP_2 PDE) in catfish cilia that is sensitive to amino acids and from IP_3-gated Ca^{2+} channels in catfish and frog.

Recent studies[19,20] have shown that different olfactory cells have different receptor proteins. It implies that each group of odor molecules is received at the corresponding olfactory cells. The electric signal transduced at olfactory cells is transmitted to the olfactory bulb, which contains about 2000 elementary bulbs named glomeruli. Surprisingly, electric signals from specified olfactory cells may converge in axons onto one or a few common target glomeruli in the bulb.

REFERENCES

1. Singer, S. J. and Nicolson, G. L. (1972). *Science*, 175, 720.
2. Ogawa, H., Sato, M. and Yamashita, S. (1968). *J. Physiol.*, 199, 223.
3. Erickson, R. P., Doetsch, G. S. and Marshall, D. A. (1965). *J. Gen. Physiol.*, 49, 247.
4. Bartoshuk, L.M. (1975). *Physiol. Behav.*, 14, 643.
5. Pfaffmann, C. (1959). In *Handbook of Physiology*, Sec. 1 *Neurophysiology*, Vol. 1, ed. J. Field., American Physiological Society, Washington, DC, p. 507.
6. Kawamura, Y. and Kare, M. R. (eds.) (1987). *Umami, A Basic Taste*. Marcel Dekker, New York.
7. Kirimura, J., Shimizu, A., Kimizuka, A., Ninomiya, T. and Katsuya, N. (1969). *J. Agr. Food Chem.*, 17, 689.

8. Birch, G. G. (1987). In *Umami, A Basic Taste*, eds. Kawamura, Y. and Kare, M. R. Marcel Dekker, New York, p. 173.
9. Lindemann, B. (1996). *Physiol. Rev.*, 76, 719.
10. Beidler, L. M. (1971). In *Handbook of Sensory Physiology*, Vol. 4, *Chemical Senses*, Part 2, *Taste*, ed. Beidler, L. M. Springer Verlag, Berlin, p. 200.
11. Akaike, N., Noma, A. and Sato, M. (1976). *J. Physiol.*, 254, 87.
12. Kurihara, K., Yoshii, K. and Kashiwayanagi, M. (1986). *Comp. Biochem. Physiol.*, 85A, 1.
13. Kinnamon, S. C. (1988). *Trends Neurosci.*, 11, 491.
14. Yamamoto, T. (1996). *Nou to Mikaku* (*Brain and the Sense of Taste*), Ch. 4. Kyoritsu Shuppan, Tokyo [in Japanese].
15. Amoore, J. E. (1962). *Proc. Sci. Sec., Toilet Goods Assn*, 37, 1.
16. Sato, T., Hirono, J. and Tonoike, M. (1992). *Sens. Mater.*, 4, 11.
17. Buck, L. and Axel, R. (1991). *Cell*, 65, 175.
18. Enomoto, S., Kashiwayanagi, M. and Kurihara, K. (1991). *Biochem. Biophys. Acta*, 1062, 7.
19. Mori, K. and Yoshihara, Y. (1995). *Progr. Neurobiol.*, 45, 585.
20. Buck, L. (1996). *Annu. Rev. Neurosci.*, 19, 517.

3

Biomimetic membrane devices

This chapter describes biomimetic membrane devices, which were developed by utilizing a self-assembled characteristic of biomolecules. First, some typical self-organized phenomena such as rhythm and pattern formation are described, because biomimetic membrane devices work far from equilibrium. It is shown that lipid membranes can show action potential, which is a transient all-or-non change of membrane potential elicited under nonequilibrium conditions with applied stimuli such as an electric current, i.e., this membrane can be regarded as an excitable model membrane or an artificial nerve membrane similar to natural nerve membranes. Since the frequency of self-sustained oscillations increases with increasing DC electric current and/or pressure, this membrane is a kind of DC–AC converter made of organic materials, or a chemo-mechanical receptor. It can also be used as a memory element or switching element with analog-to-digital function. Anesthetic substances, local anesthetics and alcohol, can stop the excitation in this biomimetic membrane, which is how anesthesia acts in biological systems. The oscillations are also affected largely by chemical substances producing taste.

Furthermore, this membrane responds to taste substances by changes in membrane electric potential and electric resistance under conditions in which oscillations do not occur. Membranes made of different types of lipid respond to chemicals in different ways. This was the first indication that lipid membranes could be used as transducers to transform taste information to electric signals. Lipid immobilized membranes are a prototype for biomimetic, biomolecular devices to reproduce the sense of taste.

3.1 Self-organization appearing far from equilibrium

Spatio-temporal patterns are seen in chemical reactions when these are coupled with diffusion. Spiral or circular chemical waves are spontaneously produced in Belouzov–Zhabotinsky (B–Z) reactions.[1] Figure 3.1 shows an

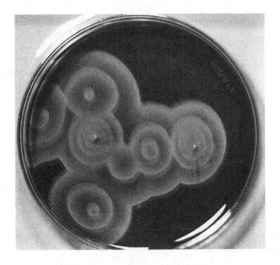

Figure 3.1. Circular waves in B–Z reactions.

example of circular waves in B–Z reactions. Self-sustained oscillations of chemical products (Ce^{4+} and Ce^{3+}) in the homogeneous space were found by Belousov when $KBrO_3$ and Ce^{4+} were added to citric acid in dilute sulfuric acid.[2] Zhabotinsky and his co-workers confirmed this result and further observed the spatial periodic structure and propagation of concentration waves.[3]

The B–Z reaction is considered to comprise 10 elementary reaction processes.[4] The simplified scheme of the B–Z reaction is illustrated in Fig. 3.2,[4,5] where the chemical reaction A + B → C is represented by

Using Fig. 3.2, let us explain the mechanism of self-sustained oscillation in the Belousov system. Chemicals BrO_3^-, $BrCH(COOH)_2$ and $CH_2(COOH)_2$ can be controlled externally. When there is a plenty of Br^-, the reactions (5) and (10) will not proceed. Instead, reaction (2) occurs and then $HBrO_2$ is consumed (first step). However, reaction (5) starts if $HBrO_2$ is consumed sufficiently in reaction (2). In addition, the reaction has a positive feedback property and hence Ce^{4+} is produced rapidly (second step). As the third step, reaction (10) starts to occur because of excess Ce^{4+}. Consequently, Ce^{3+} and Br^- increase. This allows recovery of the first step. In this way, the above process can be repeated. We can observe this periodic process as a change in

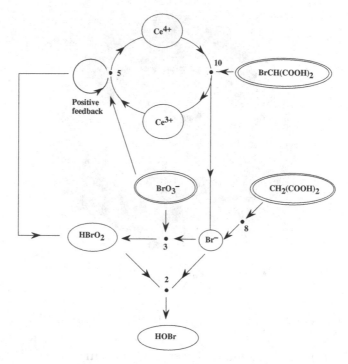

Figure 3.2. Scheme of B–Z reactions.

color associated with the presence of Ce^{4+} or Ce^{3+}. This occurs when the solution is stirred and becomes homogeneous. If the solution is not stirred, we can observe the spatial pattern by almost the same mechanism as in Fig. 3.1. These types of ordered state appear through some intrinsic nonlinear properties of a system far from equilibrium.

In biological systems, a nerve fiber exhibits excitation (i.e., a repetitive firing) to process an information transformation. This phenomenon appears as a result of self-organization realizing far from equilibrium. Similar electrical events can be found in more dynamic phenomena in biological systems, for example in growth, differentiation and morphogenesis.[6–8]

Study of simple artificial systems may be useful for understanding self-organization phenomena in biological systems and also creates the possibility of developing a novel advanced device.

3.2 Phase transition of artificial lipid membranes

A membrane constructed from a porous filter (Millipore Corp, cellulose ester, average pore size 5 or 8 μm) impregnated with dioleyl phosphate (DOPH) changes its electrical characteristics according to variations in the concentration of a salt such as NaCl or KCl.[9-13] DOPH is a synthetic lipid obtained by hydrolyzing oleyl alcohol with phosphorus oxychloride and was first made by Kobatake and Yoshida. Its chemical structure includes two hydrophobic hydrocarbon chains and one hydrophilic phosphate group:

$$CH_3(CH_2)_7CH=CH(CH_2)_7CH_2O \diagdown \diagup O$$
$$P$$
$$CH_3(CH_2)_7CH=CH(CH_2)_7CH_2O \diagup \diagdown OH$$

DOPH has a similar structure to the typical phospholipid molecules in biological systems. Compared with phosphatidylcholine, a natural phospholipid (see p. 27), we can see that DOPH is obtained by adopting part of the phosphate group and removing the ammonium part of the choline group. DOPH is a yellow, liquid oil at room temperature.

The membrane was placed between two cells with equimolar ion concentrations, as shown in Fig. 3.3. When the ion concentration is below about 30 mM, the membrane shows low electric capacitance, of the order of $10 \, nF/cm^2$, and electric resistance as high as several megaohms per square centimeter. Above 30 mM, however, these membrane electrical characteristics change to approximately $300 \, nF/cm^2$ and a few hundred kiloohms per square centimeter, respectively. The measured electric capacitance is shown in Fig. 3.4.

The membrane capacitance shows a sigmoidal relaxation after an abrupt increase in ion concentration, but an exponential relaxation after a decrease (Fig. 3.5). The speed is as slow as several minutes or more for the increasing process, although it is relatively fast for the decreasing case. Consequently, a hysteresis loop is obtained for one cycle of increasing and decreasing processes; the loop belongs to a different type from the usual hysteresis originating from a first-order phase transition in an equilibrium system[14] or a hard transition (sometimes called an inverted-type or supercritical bifurcation[1]) in nonequilibrium systems (Fig. 3.6).[15] The relaxation process from the low ion concentration to the high concentration becomes slower when the quantity of DOPH adsorbed onto the membrane filter is increased. The process took a few hours for a membrane with $1-2 \, mg/cm^2$ DOPH adsorbed but over 24 h for a membrane with over $3 \, mg/cm^2$ DOPH adsorbed.

A hysteresis loop of the membrane electric capacitance was obtained by repeating the measurements after the passage of time interval $\Delta\tau$ from each abrupt step-by-step change of ion concentration.[13] In Fig. 3.4, one hysteresis loop obtained for $\Delta\tau = 5 \, min$ and a reversible equilibrium curve obtained for

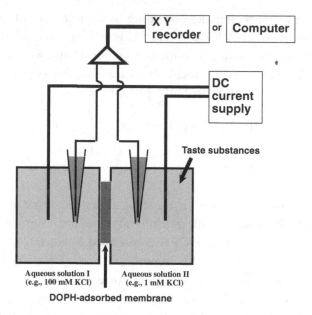

Figure 3.3. Experimental setup for measuring the electrical characteristics of the DOPH-adsorbed membrane. The area of about $1\,cm^2$ of the membrane is in contact with two aqueous solutions, which are changed according to the experimental purpose. For example, in the investigation of phase-transition characteristics, they were chosen to have equimolar concentrations, while they were 1 mM (or 5 mM) KCl and 100 mM KCl solutions in the study of short-period oscillations (Section 3.3.1). The electric potential across the membrane was measured using a pair of KCl-saturated Ag/AgCl electrodes by taking its origin at the 1 mM KCl side. See also Fig. 6.15. The electric oscillations were induced by application of DC electric current using a pair of Ag/AgCl electrodes. When measuring the membrane electric capacitance, the Ag/AgCl electrode was replaced by a plate-like Pt/Pt electrode, which is adequate for AC measurements.

$\Delta\tau > 30$ min are shown. At low ion concentrations, the membrane shows the low electric capacitance; however, it shows the high electric capacitance at high ion concentrations. The hysteresis appears with high scanning rate but shrinks to one reversible equilibrium curve with a low scanning rate. The solid and dashed lines imply the theoretical results explained in Section 3.4.1.

Any hysteresis phenomenon is essentially time dependent, because any system cannot stay in a metastable state eternally. The dynamic aspect of the hysteresis phenomenon has been investigated extensively as a result of recent progress in nonequilibrium thermodynamics. In ferromagnetic materials and type II superconductors containing many pinning centers, however, the decay of the hysteresis loop of magnetization with time is too slow to be observed under usual experimental conditions, and hence these materials are not very suitable for an investigation of the dynamic profile of the hysteresis.

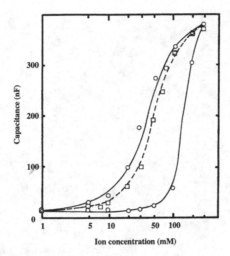

Figure 3.4. Change in the electric capacitance of the DOPH-adsorbed membrane with NaCl concentration.[13] Experimental data are shown by ○ and □, theoretical results shown by solid and dashed lines, respectively.

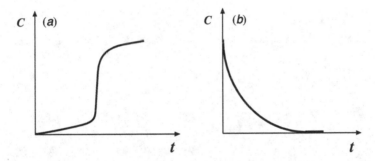

Figure 3.5. Schematic illustration of relaxation of membrane capacitance showing sigmoid change after increasing ion concentration (*a*) and exponential change after decreasing concentration (*b*).

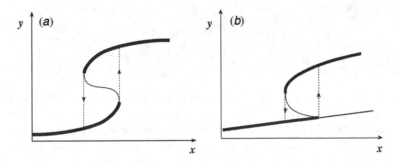

Figure 3.6. First-order phase transition in an equilibrium system (*a*) and a hard transition occurring far from equilibrium (*b*).

(a)

(b)

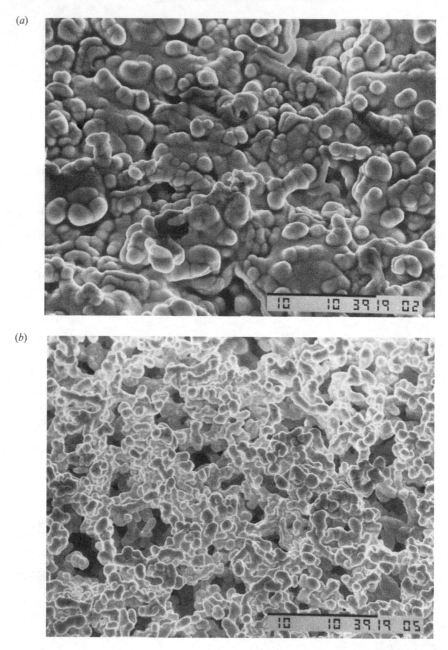

Figure 3.7. The surface structure of the DOPH-adsorbed membrane.[16] The bar at the bottom right indicates 10 µm. In 1 mM KCl (*a*), the surface is covered entirely with oil droplets; in 100 mM KCl (*b*), pores of about 5 µm appear as dark areas. Random aggregates of oil droplets occupy the pores and the surface of the filter in (*a*), and hence the electric resistance becomes high at low KCl concentrations. Formation of tightly packed multilayers causes void spaces inside pores in (*b*), resulting in low membrane electric resistance at high KCl concentrations.

For a hysteresis of the order parameter related to the first-order phase transition in equilibrium systems or a hard transition appearing far from equilibrium, however, a dynamic profile has been effectively investigated, though few detailed data have been reported on the decay of the hysteresis loop.

The electrical characteristics of the DOPH-adsorbed membrane can be explained by assuming that DOPH molecules make phase transitions among three phases composed of oil droplets at low ion concentrations, spherical micelles and multi- or bilayer leaflets at high concentrations.[9-13] The oil droplet is a loosely packed hydrophobic phase of randomly aggregated DOPH; the multi- or bilayer is a tightly packed hydrophilic phase. Spherical micelles, including inverted micelles, are formed from about 100 lipid molecules or more.

Figure 3.7 shows a photograph of the surface structure of the membrane.[16] At 1 mM KCl, the oil droplets cover the entire surface. At 100 mM KCl, pores of about 5 μm size can be seen as dark areas. These results agree with the high membrane electric resistance seen at low concentrations and low resistance at high concentrations. Furthermore, DOPH molecules can form a bilayer lipid membrane (black lipid membrane[17]) at high ion concentrations;[9] this fact supports the above phase transition. Stabilization of a multi- or bilayer structure at high ion concentrations can be shown by an electrochemical theory (see Section 3.4.1).

3.3 Excitability: self-sustained oscillations

Oscillatory behaviors have been reported for several simple membranes. One is a glass filter placed under electrochemical and hydrodynamic gradients. The oscillation of membrane potential was found by Teorell[18] and explained theoretically by Kobatake.[19] The others may be systems containing lipids such as an oil membrane,[20] a bilayer lipid membrane[21] and an oil/water interface.[22,23] These membranes (or interface) can reproduce excitability as found in nerve membranes, although these systems do not contain proteins but only lipids.

3.3.1 Five types of oscillation in the DOPH-adsorbed membrane

The DOPH-adsorbed membrane also shows excitability under non-equilibrium conditions. Figure 3.8 summarizes the oscillations. The first is a long-period oscillation, which appears when the membrane is placed between

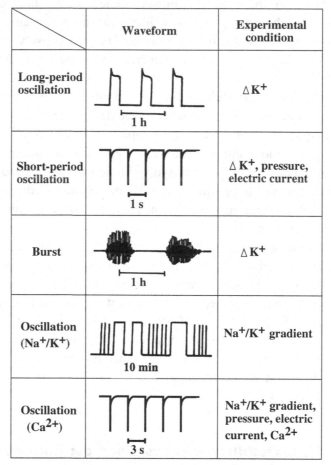

	Waveform	Experimental condition
Long-period oscillation	1 h	ΔK^+
Short-period oscillation	1 s	ΔK^+, pressure, electric current
Burst	1 h	ΔK^+
Oscillation (Na^+/K^+)	10 min	Na^+/K^+ gradient
Oscillation (Ca^{2+})	3 s	Na^+/K^+ gradient, pressure, electric current, Ca^{2+}

Figure 3.8. Five kinds of oscillation appearing in the DOPH-adsorbed membrane.

solutions of low and high ion concentrations, e.g., 5 mM and 100 mM KCl, without applied DC electric current.[11,12,24–26] The origin of the electric potential was taken at the 5 mM KCl side. The period is as long as several minutes to a few hours, the waveform exhibiting the spike form. The period and the width of the spike become longer as the amount of DOPH adsorbed into a porous filter is increased.[24] Since the side with the higher ion concentration is electrically negative, the DOPH membrane is permeable to cations; in this case, the membrane electric resistance is not so large, and hence K^+ diffuse from the 100 mM to the 5 mM solution. This occurs because the negative charge of the phosphate group of a DOPH molecule is dissociated near a neutral pH. This oscillation appears by coupling between the phase transition of DOPH and ion flow across the membrane.[25]

The second oscillation is a short-period oscillation, which is induced by application of a pressure gradient and DC electric current under an ion concentration gradient.[25–29] The period is 1 s or so and the waveform is a relaxation oscillation showing exponential decay. The oscillation continues for several tens of minutes or sometimes over 1 h. It can be induced by increasing the applied DC electric current or pressure difference, as in Fig. 3.9. The oscillation appeared beyond 0.32 μA and its frequency increased linearly with increasing electric current. Since the input is the DC electric current and the output is the AC electric voltage, the membrane can be regarded as a DC–AC converter. The pressure difference also increases the frequency of the oscillation, and hence the membrane acts as a mechanical receptor to detect pressure.[28]

The third oscillation is a burst-type oscillation composed of fast and slow oscillations appearing under the ion concentration gradient. This oscillation occurs through nonlinear coupling between the short-period oscillation by DOPH molecules attached to the surface region of a filter and a long-period oscillation by DOPH adsorbed within a pore.[25]

The fourth is an oscillation appearing under a 100 mM NaCl/100 mM KCl gradient.[30] The membrane potential is negative at the KCl side, and hence the DOPH membrane is permeable to K^+ more than to Na^+. This is quite similar to the situation in biological membranes. The oscillation is more sensitive to K^+ than Na^+, since the frequency is increased more with an increase in KCl concentration.

The fifth is an oscillation occurring in the presence of Ca^{2+}.[31] The oscillation appearing under the 100 mM NaCl/100 mM KCl gradient is considerably affected by application of Ca^{2+}, which induces the state of high membrane electric resistance. The waveform becomes very different from the above

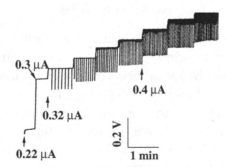

Figure 3.9. Short-period oscillation. The pressure of 29.5 cmH$_2$O and the DC electric current are applied on the DOPH-adsorbed membrane from the 100 mM KCl to the 5 mM KCl side.

fourth oscillation in the absence of Ca^{2+}. In biological systems, Ca^{2+} plays an essential role in excitation, which originates from opening/closing of ion channels. This membrane can be used as an excitable element regulated by Ca^{2+}.

The part showing the oscillation can be considered as one pore or several close pores behaving cooperatively. One reason is that we have often observed coexistent appearance of two kinds of oscillation with different amplitudes and periods (see, for example Fig. 3.18, below). Another reason is that estimation of electric resistance of an open pore leads to a good explanation of the observed data. Using the conductivity of 100 mM KCl ($150\,cm^2/\Omega$ per equiv.), an open pore with, for example, a radius of 2.5 μm and length 100 μm has an electric resistance of about 3.4 MΩ. Since the membrane resistance R_m is several megaohms or more at a high-resistance state, we get 5×10^6 MΩ as the electric resistance of one "closed" pore from 10^6 pores by adopting, for example, a value of 5 MΩ as R_m. It implies that the electric resistance of one pore can change from 5×10^6 MΩ to 3.4 MΩ. In this "open" state, R_m decreases to $5 \times 3.4/(5 + 3.4) = 2.0$ MΩ. We can conclude that the membrane resistance R_m can change between 5 and 2 MΩ while one pore is closed and open. This result is consistent with the observed change of membrane electric resistance.[31]

3.3.2 Observation of oscillation in a single-hole membrane

Two types of single-hole cell were used in the present experiments.[32] One is for microscopic observation with the optical axis parallel to the axis of the hole, the other for microscopic observation with the optical axis perpendicular to the hole axis.

Axial (type A) cell

The thickness of the cell was approximately 3 mm, which is the limiting thickness at present for microscopic observation. A filter with a single hole was placed between two cells, one of them containing 100 mM KCl solution and the other 5 mM KCl solution. The single-hole filter was prepared in the following way. The filter was varnished and left for one day at room temperature. This filter became hydrophobic and all the pores in the film were blocked. Then a new hole was created with a heated tungsten needle. The diameter of the tungsten needle sharpened by electrolytic polishing was controlled by the current and time of polishing. In this experiment, the diameter of the hole was 50 μm.

Transverse (type B) cell

To observe transversally the oscillatory phenomenon in a hole, a thin cell was fabricated by means of semiconductor technology. A $55 \times 9 \ \mu m^2$ single-hole membrane was made; DOPH and KCl solutions were introduced through the different paths into the cell.

Observations

Axial observation of the hole using a type A cell showed the sequence of changes shown in Fig. 3.10. The hole is almost homogeneous and dark in (*a*). The central part of the hole becomes suddenly bright (less than 0.33 s) in (*b*). This is followed by the relatively slow change from the border (1 s) in (*c*) and the hole reverts to its original state in (*d*). Such a phenomenon was observed periodically.

Figure 3.11 shows the sequence of transverse observations of the oscillation phenomenon. The interface between the 100 mM KCl solution and DOPH is pushed down, while the interface between the DOPH layer and the 5 mM KCl solution remains unchanged in (*a*) and (*b*). The DOPH layer blows up in (*c*) and (*d*). Then the hole begins to be repaired in (*d*). This series of changes was observed periodically.

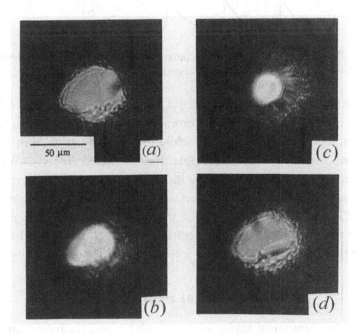

Figure 3.10. Axial observation of a single hole in a membrane to show the oscillation.[32]

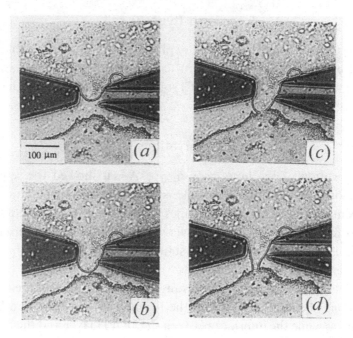

Figure 3.11. Transverse observation of a single hole in a membrane to show the oscillation.[32]

Figure 3.12. The oscillation of the electric potential measured simultaneously with the observation shown in Fig. 3.10.[32]

The membrane potential across the DOPH layer, associated with this oscillation, is shown in Fig. 3.12. Comparison with Fig. 3.10 implies that the membrane potential remains low while the "DOPH gate" is open. Then it was closed, and the membrane potential increased linearly. When the DOPH layer blew up, the potential suddenly fell to the original level.

These observations clearly show what was taking place at the two interfaces of the DOPH layer with different KCl solutions. The open/close process is repeated to result in the membrane electric oscillation. The DOPH-adsorbed membrane is excitable and acts as an ion-concentration-sensitive ion gate.

3.4 Theoretical explanation

In this section, theoretical explanations for the phase transition and short-period electric oscillation of the DOPH-adsorbed membrane are discussed.

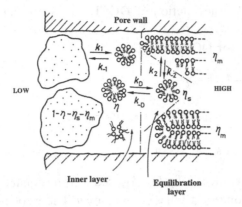

Figure 3.13. Theoretical model for the conformational states of DOPH assemblies in a pore when the membrane is placed between the low and high ion concentrations.[29] An oil droplet is shown by the amorphous shape with dots.

Figure 3.13 illustrates the conformational states of DOPH assemblies in a pore when the membrane is placed between the low and high ion concentrations.[24–26,29] The electric oscillation can occur in one pore, as seen in the above estimate of membrane electric resistance. Furthermore, Fig. 3.12 shows that a DOPH-adsorbed film with a single minute pore can generate electric oscillations.[32,33] In Fig. 3.13, a DOPH molecule is illustrated by one circle with two zigzags. It is important that the multilayer structure is formed in parallel with the wall of a pore; then the void space is produced in the pore because of the high packing density of multilayers. This void space causes ion permeation through the membrane.

3.4.1 Phase transition

Let us then give kinetic equations for the phase transition of DOPH molecules. We define η as the fraction of DOPH molecules in micelles in the inner bulk layer far from the pore wall, η_m as that of multilayers near and parallel to the wall and η_s as (spherical) micelles in an equilibration layer, which is in contact with the above two phases:

$$\eta = [\text{micelles in inner bulk layer}]/N,$$

$$\eta_m = [\text{multilayers}]/N, \tag{3.1}$$

$$\eta_s = [\text{micelles in equilibration layer}]/N,$$

where [M] denotes the number of DOPH molecules with the structure specified by M, and N is the total number of DOPH related to the phase transition (and the oscillation). The fraction of DOPH in oil droplets corresponds to $(1 - \eta - \eta_s - \eta_m)$.

Kinetic equations for DOPH are given by

$$\frac{d\eta}{dt} = k_1(1 - \eta - \eta_s - \eta_m)n - k_{-1}\eta + k_{-D}\eta_s - k_D\eta,$$

$$\frac{d\eta_s}{dt} = -k_{-D}\eta_s + k_D\eta + k_{-2}\eta_m - k_2\eta_s, \quad (3.2)$$

$$\frac{d\eta_m}{dt} = -k_{-2}\eta_m + k_2\eta_s,$$

where the coefficients $k_1, k_{-1}, k_2, k_{-2}, k_D$ and k_{-D} designate the relevant rate constants. The ion concentration is denoted by n. These coefficients depend on the number of DOPH molecules because the phase transition is essentially a sequential reaction between oil droplets and multilayers. The coefficient k_{-2} (or k_2) depends on the ion concentration n, because the electric double layer is formed around the charged phosphate group of the DOPH molecule.[13]

For the transformation from oil droplets to small spherical micelles, the following simple reaction was assumed:

$$\text{(oil droplets)} + p(\text{monovalent cations}) \underset{k_{-1}}{\overset{k_1}{\rightleftharpoons}} (\text{micelles}), \quad (3.3)$$

where p is the parameter of the order of unity.

An exchange of DOPH micelles between the equilibration and inner layers may be mainly caused by diffusion. However, the spherical micelles cannot move freely among oil droplets when the fraction of DOPH molecules in droplets is large and that of multilayers is small, because the size of a droplet may be of a comparable order with a pore size, as seen from Fig. 3.7(a). Then the diffusion of spherical micelles is thought to be slower for smaller values of the fraction of multilayers η_m. In other words, the rate constants k_D and k_{-D} may be considered as the increasing functions of η_m. This tendency can be expressed with the aid of the relation between a diffusion constant and free volume.[34] In the present case, the free volume, v_f, where spherical micelles can move freely, is assumed to have the simple dependence on η_m given by

$$v_f \propto 1 + d_2\eta_m^r \quad (3.4)$$

with the constant parameters of d_2 and r. Since the diffusion arises from the difference between "concentrations" at different points, the rate constants k_D and k_{-D} used for expressing the diffusion will contain such a property. Therefore:

$$k_D = k_{-D}/v, \quad (3.5)$$

where k_D is given by

$$k_D = D_0 \exp[-d_1/(1 + d_2\eta_m^r)] \quad (3.6)$$

with constant parameters of D_0, d_1 and v. Among them, v is the ratio of k_D and k_{-D} and is mainly subject to the volume ratio of the equilibration layer to the inner one. This volume is much smaller than unity; hence, v will take a large value.

Now let us consider the transformation between spherical micelles and multilayers. This reaction may depend on the ion concentration because of the electrochemical energy accompanying formation of the multilayer. For DOPH molecules near neutral pH, a polar-head group has one negative charge and hence the distance between bilayers may be large, which allows the approximation that multilayers are composed of independent bilayers in contact with the water phase. Therefore, the electrochemical energy (G) for one DOPH molecule is given by (see also Section 6.6.1)[35]

$$G = k_B T \int_0^\alpha \ln\left(\frac{[H^+]}{K}\frac{\alpha}{1-\alpha}\right) d\alpha + A \int_0^\sigma V_s \, d\sigma, \qquad (3.7)$$

where the surface charge density σ is given by

$$\sigma = -(e/A)\alpha \qquad (3.8)$$

with α designating the degree of dissociation of protons given by

$$\alpha = -2Aqn^{1/2}\sinh(\phi/2), \qquad (3.9)$$

$$\phi = eV_s/k_B T. \qquad (3.10)$$

In eqs. (3.7)–(3.10), $[H^+]$ is the proton concentration far from the bilayer, K the dissociation constant of protons, A the molecular surface area of DOPH, V_s the electric potential at the surface of bilayer, e the positive elementary charge, k_B the Boltzmann constant, T the absolute temperature and q is defined by

$$q = \left(\varepsilon k_B T/2\pi e^2\right)^{1/2} \qquad (3.11)$$

with ε the dielectric permeability of water. With the aid of a high potential approximation of $e^{-\phi} \gg 1$ at $\alpha \simeq 1$ and $n \geq 1\,\mathrm{mM}$, we obtain an approximate expression for the electrochemical energy as

$$G = k_B T\{\ln([H^+]/K) - 2\ln(Aqn^{1/2}) - 2 + 4Aqn^{1/2}\}. \qquad (3.12)$$

Since the rate constant k_{-2} is proportional to $\exp G$ and the last term in the brace in eq. (3.12) is much smaller than unity, k_{-2} is given by

$$k_{-2} = k^0_{-2}/n. \qquad (3.13)$$

Fusion between spherical micelles and multilayers is cooperative to some extent because of the van der Waals attractive interaction; consequently, the reaction rates k_2 and k_{-2} will contain such a property. However, detailed functional forms of k_2 and k_{-2} scarcely affect the results, and hence k_{-2}^0 defined in eq. (3.13) and k_2 are assumed as constant.

The expression for k_{-2} in eq. (3.13) implies that formation of the multilayer is more favorable with increasing ion concentration. This result is consistent with the experimental fact that the DOPH bilayer membrane can form at high ion concentrations.[9]

The ion concentration n refers to that within the pore concerned. We consider the situation where two aqueous phases in contact with the membrane are composed of the same ion concentration, as in Fig. 3.4. If n is chosen as equal to the ion concentration in the cells bathing the membrane, the above equations can explain the phase-transition characteristics quantitatively by assuming that the membrane electric capacitance is proportional to the fraction of multilayers η_m.[13]

The result is compared with the observed data in Fig. 3.4. The observed hysteresis loop and reversible equilibrium curve are explained quantitatively. The multilayer formation is accelerated at higher ion concentrations because the electric repulsion between charged head groups of DOPH is weakened by cations (electric screening effect). In this way, the present theoretical model can describe the relaxation process of phase transition of DOPH assemblies.

3.4.2 Self-sustained oscillation

When the membrane is placed under the ion concentration gradient, the ion concentration n (i.e., the K^+ concentration) within the pore can be changed with the state of DOPH assemblies, which affects the ion permeability. The membrane electric resistance is high in the oil droplets at low ion concentrations, whereas it is low in the multilayer state at high concentrations. We assume the following simple equation for n:[23–25,29]

$$\frac{dn}{dt} = D(\hat{n} - n), \tag{3.14}$$

where D corresponds to the diffusion constant of K^+ and $\hat{n}$ is the stationary concentration dependent on the state of DOPH given by

$$\hat{n} = \frac{N_0[B_1 + (\zeta/\zeta_c)^\gamma][1 + B_2(\eta_m/\eta_{mc})^\lambda]}{[1 + (\zeta/\zeta_c)^\gamma][1 + (\eta_m/\eta_{mc})^\lambda]} + dQ, \tag{3.15}$$

where ζ is the total micelle fraction:

$$\zeta = \eta + \eta_s \tag{3.16}$$

with N_0, B_1 and B_2 the numerical parameters for the magnitude of n. The function $\hat{n}$ with parameters $\zeta_c, \eta_{mc}, \gamma$ and λ expresses that the multi-bilayer formation causes release of K^+ while micelles accumulate K^+. The quantity Q in the second term with the constant d in eq. (3.15) is the electric charge accumulated inside the membrane from the applied electric current I; it is expressed by

$$Q = CV_m,$$ (3.17)

where V_m is the membrane electric potential given by

$$V_m = \frac{R}{R_0 + R}(R_0 I - E_0) - \frac{R}{R_0 + R}E.$$ (3.18)

The measured value of V_m can be expressed by an equivalent electric circuit for the membrane, as shown in Fig. 3.14. The terms C, R and E are the electric elements of the region where the oscillation occurs, while C_0, R_0 and E_0 refer to the remaining region independent of the oscillation. The values of C, R and E vary temporarily with the change in the conformational structure of DOPH assemblies. The change in the voltage across the condenser C is much faster than the rate of conformational change in DOPH.[8] Because of small values of C_e and R_e (the electric capacitance and resistance, respectively, in the external aqueous solution), we can safely approximate the membrane potential V_m by eq. (3.18).

For simplicity, we assume $E = 0$ and R has the functional form:

$$R = R_h + R'\left(\frac{1}{\eta_m} - 1\right),$$ (3.19)

because R takes low and high values at high and low ion concentrations, respectively, which correspond to high η_m (nearly unity) and low η_m (nearly

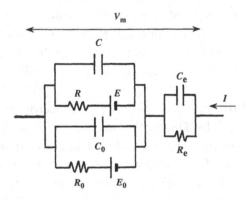

Figure 3.14. Equivalent electric circuit. The electric capacitance and resistance in the external aqueous solution are denoted by C_e and R_e, respectively. The electric characteristics of the membrane are expressed by the electric circuit composed of the oscillating part (C, R, E) and nonoscillating part (C_0, R_0, E_0).

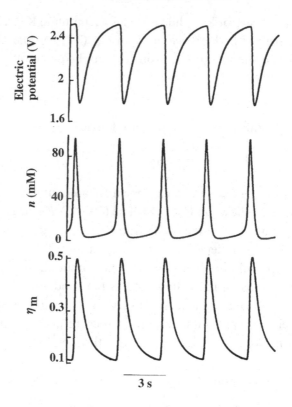

Figure 3.15. The calculated results for the short-period oscillation.

zero). Equation (3.19) is reduced to R_h for $\eta_m \simeq 1$, while it becomes $R'/\eta_m (\gg R_h)$ for $\eta_m \simeq 0$.

Basic equations for describing the oscillations of η, η_s, η_m and n are eqs. (3.2) and (3.14), and the membrane electric potential V_m is described by eq. (3.18) with eq. (3.19) for R. The calculated results are shown in Fig. 3.15. The waveform of membrane potential reproduces well the observed waveform in Fig. 3.9. The ion concentration and the fraction of multilayers also oscillate in phase.

Dependencies of the frequency on the applied electric current and pressure are also shown in Fig. 3.16. The oscillation appears beyond the critical applied current, and its frequency increases almost linearly. It increases with the applied pressure. These facts agree with the observed data.

Figure 3.17 illustrates the mechanism of the oscillation. The region considered now is a thin region of about 1 μm length located at the lower-ion-concentration side. Most of the remaining part (~100 μm length) of the filter paper is not related to the oscillation, since the conformational states of DOPH are not changed during the oscillation because of the high ion

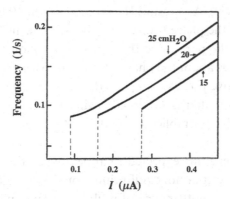

Figure 3.16. The calculated result of dependencies of the frequency on the applied electric current and pressure.

concentration. First, let us assume that the pore is occupied by oil droplets at the lower-concentration side (Fig. 3.17(*a*)). Then, the membrane electric resistance should be so high that ions cannot flow across the membrane. As a result, K$^+$ accumulates within the pore. Increasing K$^+$ concentration leads to the phase transition of DOPH molecules from oil droplets to micelles and

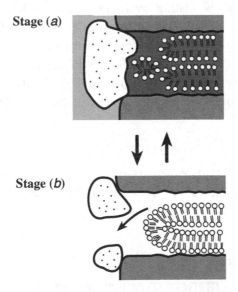

Figure 3.17. Mechanism of the oscillation. The left and right sides of the membrane are 1 mM and 100 mM KCl solutions, respectively. The ion concentrations of the left-hand solution and in a pore are reflected by shading. (*a*) The pore is occupied by droplets at the lower-concentration side. (*b*) Phase transition occurs in the oil droplet as a consequence of the accumulation of K$^+$ within the pore. The arrow in (*b*) implies quick outflow of ions from the pore.

multilayers (Fig. 3.17(*b*)). As a result, K^+ is released from the pore because the pore opens following the phase transition. The decrease in K^+ concentration should occur within the pore since the concerned region touches the lower concentration; it brings about the phase transition to oil droplets. Stage (*a*) is recovered and the above process is repeated. The self-sustained oscillation of membrane electric potential is observed owing to this repeated conformational change in DOPH assemblies associated with ion release and accumulation.

The oscillation is generally expected to appear when the system showing a phase transition is placed in nonequilibrium conditions. Electric oscillations have been observed for another type of multi- or bilayer membrane made of synthetic cationic lipid showing an ordered–disordered (ordered–fluid) phase transition with temperature.[36] A mechanism of oscillation can be theoretically derived[37–39] and shown to be essentially the same as that of the DOPH-adsorbed membrane; the oscillation is brought about by coupling between ion flow across the membrane and the phase transition, which is affected by ionic circumstance. It is reasonable to consider that self-sustained electric oscillations will appear also in other many systems containing only lipids.

The temporal order appearing in artificial lipid membranes can be easily controlled. If we define the resting level and the excited level as two states of the membrane, we can utilize the membrane as a switching element. Coupled switching elements would produce a spatio-temporal order, as tried using a reaction-diffusion system.[40] Even in the DOPH-adsorbed membrane, the coupled oscillation is found,[41] as shown in Fig. 3.18(*a*). This oscillation may arise from two pores interacting with each other through electric current

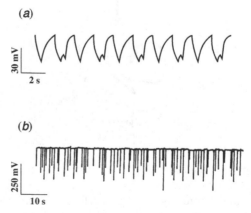

Figure 3.18. Mutually entrained oscillation (*a*) and random oscillation composed of several independent oscillations (*b*).

or ion concentration. In most cases, however, several oscillations seem to appear at the same time independently and randomly, which occurs under strong applied DC electric current (see Fig. 3.18(*b*)).

3.5 Effects of anesthetics and taste substances on excitability of the lipid membrane

Excitability is one of the most important properties of biological membranes. The excitation of membranes involves transient depolarization (i.e., action potential) or repetitive firing (i.e., oscillation) of the membrane potential. As seen in the previous section, a membrane made of lipids can show a similar excitability. The present section examines the effects of anesthetics and of chemical substances that produce bitterness on the excitable lipid membrane; these chemicals have large effects on the excitability of the DOPH-adsorbed membrane.

3.5.1 Effect of anesthetics

Local anesthetics such as tetracaine act in biological systems by suppressing the action potential.[42] Does the same situation occur in artificial lipid membranes? The site of action of anesthetics is considered to be nerve cell membranes. The mode of action on the membrane is still unclear despite numerous investigations. One hypothesis suggested that the site of anesthetic action would be hydrophobic in nature to interact with hydrocarbon parts of lipids because of a high correlation between anesthetic potency and the solubility of the anesthetic in olive oil.[43–45]

Figure 3.19 shows suppression of the short-period oscillation of the DOPH-adsorbed membrane induced by tetracaine.[46] The experimental situation is the same as that in Fig. 3.3; the membrane was placed between a 100 mM and 1 mM KCl solution and tetracaine was added to the latter. The 1 mM KCl side can be regarded as the outside of a biological cell from comparison with a real system. The oscillation amplitude decreased with 0.3 mM tetracaine and then stopped, i.e., the excitation is inhibited by tetracaine. Therefore, this lipid membrane is capable of reproducing a characteristic property of nerves. Of course, the mechanism is different in the artificial lipid membrane and the nerve membrane because channel proteins play a role in excitation in biological systems. However, there is a high possibility that the lipid part of biological membranes plays a role in receiving anesthetic substances, as also shown by other experiments, described below, using the DOPH-adsorbed membrane.

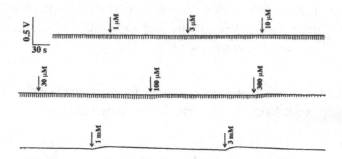

Figure 3.19. Suppression of the short-period oscillation by tetracaine.[46] Experimental
conditions: $30 \, \text{cmH}_2\text{O}$ and $0.10 \, \mu\text{A}$ were imposed on the membrane.

Anesthetics such as procaine and lidocaine (lignocaine) with less potent
action than tetracaine induced irregular and aperiodic oscillations. In
biological systems, some reports have been made of aperiodic or chaotic
oscillations in information processing such as olfactory recognition.[47,48] It
may be interesting to pursue whether aperiodic oscillations appear in nerve
membranes after application of anesthetics.

The present results on local anesthetics are the first demonstration of
anesthetic actions on the excitability of lipid membranes. The effect of *n*-
alkanols on anesthetic potency is well known; potency increases with increasing chain length of alkanols until potency is suddenly lost at approximately 12
carbon atoms.[49,50] This effect can be reproduced using the DOPH-adsorbed
membrane.[51] These results might clarify the mechanism of anesthetic action.

It may be possible to use the lipid membrane for the detection or measurement of anesthetics. This is because anesthetics in general have hydrophobic
(and also hydrophilic) properties, which enable anesthetics to adsorb onto
membranes made of amphiphilic lipid molecules. As a result, the membrane
potential, membrane electric resistance and the excitability of the membrane
are substantially altered.

3.5.2 Effect of bitter substances

As seen above, chemical substances with anesthetic potency affect the electric
oscillation of the lipid membrane. This implies that information regarding
chemical substances can be obtained from changes in electric oscillations. It
is, therefore, expected that nonequilibrium situations of dynamic changes
including electric oscillations in lipid membranes might be utilized for
chemical sensing. We can expect construction of an artificial sensing system

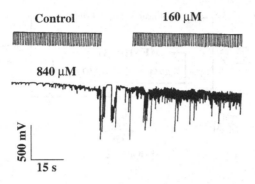

Figure 3.20. Effect of quinine to show how strong bitterness alters electric oscillations.[52]

for taste to use lipid membranes. As one of the first prototypes, a model membrane made of DOPH molecules may be effective.

Bitter substances were added to the 1 mM KCl solution in the same way as anesthetics. Figure 3.20 shows the effect of quinine, which produces strong bitterness, on the electric oscillation.[52] While the oscillation is initially regular with a frequency of approximately 0.5 Hz, it becomes irregular when several hundred micromoles of quinine is applied. This type of irregular oscillation was usually induced by employing strong bitter substances such as strychinine, quinine and nicotine. Substances with weak bitternesss such as caffeine and theobromine did not induce pronounced changes in the oscillation.

As discussed in the following section and Section 7.2 (and theoretically explained in Section 6.6.1), strong bitter substances are adsorbed into the lipid membrane because of its hydrophobicity. This affects the assembly of DOPH molecules within pores; hence, the oscillation, which results from repeated conformational changes of lipid assemblies, is much affected.

Other chemical substances to show saltiness, sourness, sweetness and umami taste also affected the electric oscillation,[16,29] although the changes were smaller than with bitter substances such as quinine. The frequency and amplitude changed in different ways with different taste qualities. Consequently, this lipid membrane can be considered as a prototype for a taste sensor. If we analyze the changes in waveform as well as frequency, we can derive information of taste sensation. Figure 3.21 illustrates a basic concept of a taste sensor utilizing oscillations.[53] As detailed in Chapters 6 and 7, however, utilization of changes in membrane potential of static (strictly speaking, "stationary") state lipid membranes can be considered as favorable at present because of the response stability and feasibility of data processing.

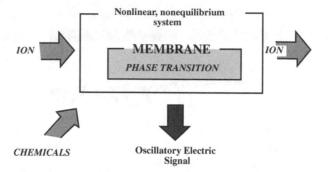

Figure 3.21. A taste sensor utilizing a self-oscillation in an artificial membrane system that works far from equilibrium.

3.5.3 Chaos in excitable lipid membranes

When a membrane was placed between two cells with equimolar KCl concentration (1 mM) and a small DC current was applied, the oscillation seen was rather regular and had a few frequency modes. Increasing the applied electric current disturbed the oscillation gradually and at last induced fully developed aperiodic oscillations.[54] The oscillation occurring after application of 60 nA had a correlation dimension converging to about 5.1, which may imply the aperiodic oscillation is chaotic with high dimensionality. The correlation dimension represents a measure of the fractal dimensionality of the waveform.

Addition of quinine to this chaotic state drastically changed the correlation dimension. At low concentrations of quinine, the dimension increased a little; it then decreased above 0.3 mM, and 10 mM quinine stopped the oscillation. Nicotine increased the dimension above 0.1 mM and then suddenly decreased it above 10 mM.

The effect of HCl (sourness) on the correlation dimension was similar to that of quinine. The dimensional changes, however, accompanied the gradual decrease of α (the slope of the power spectrum, $f^{-\alpha}$), and hence the effect was different from that of quinine, which had little effect on the value of α.

By comparison, the effects of NaCl (saltiness), sucrose (sweetness) and MSG (umami taste) on the dimension were small. NaCl and sucrose caused no change in the spectrum distribution, while MSG apparently decreased the value of α at concentrations above 30 mM. The waveform following addition of MSG was unique compared with those found from addition of other chemicals.

Application of a DC electric current caused the chaotic oscillation described above. However, the membrane exhibited a self-sustained regularly periodic oscillation of the membrane potential in the presence of an ion-concentration difference, pressure difference and DC electric current across the membrane. What happens when an AC electric current was superimposed on the DC current? Various responses of the membrane potential were studied under the periodic stimulation of sinusoidal electric current whose parameters were the frequency and amplitude.[55]

The frequency and amplitude of the applied stimulation strongly affected the frequency of the self-sustained oscillation. In the entrained state, the frequency of the oscillation was fixed at some rational number times the frequency of the applied sinusoidal stimulation. Each $m:n$-entrainment phenomenon was observed when the frequency of the applied sinusoidal electric current was near m/n (m and n are integers) times the original frequency of self-sustained oscillation without stimulation.

Quasiperiodic oscillations were found when the parameters (amplitude and frequency) of the sinusoidal stimulation were set near the $m:n$-entrained regime, i.e., near the synchronized–desynchronized phase boundary. In addition, the oscillation became chaotic when the parameters of the sinusoidal stimulation were set near the midpoints between two different entrained regimes. In the return map constructed from the firing interval (i.e., the period of self-sustained oscillation), the points on the map are tied on a distorted circle, as shown in Fig. 3.22(a),[56] which is called an attractor characteristic of chaos.

We investigated the changes in the chaotic state under AC electric current with application of four types of chemical producing four different tastes. As depicted in Fig. 3.22 (with the example of quinine), shift and deformation of the attractor by application of taste substances were observed. Quinine deformed the attractor by revolving it around the axis illustrated in Fig. 3.22(a). Each taste substance caused different changes of the attractor. As for the mean frequency shift of the attractor, quinine and HCl shifted the attractor to the high-frequency side, while NaCl shifted it to the low-frequency side. When sucrose was added, the attractor retained almost the same mean frequency.

HCl changed the structure of the attractor, i.e., the part in which the sampling points are concentrated moved to another position. NaCl made the attractor roundish. Sucrose caused the attractor to revolve around one axis and deformed the shape of the attractor.

The change in the structure of the attractor with application of the taste substances was large. The chaotic state responded sensitively to taste substances. The structural deformation of the chaotic attractor was dependent on the kind of taste substance. The attractor deformation may be induced by the interaction between taste substances and lipid molecules in the membrane.

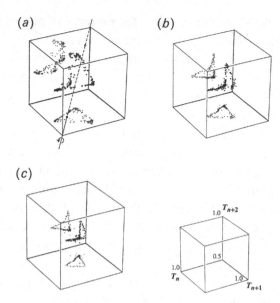

Figure 3.22. Effect of quinine on the chaotic state:[56] (*a*) no quinine, (*b*) 0.1 mM and (*c*) 1 mM. T_n is the normalized value of the firing interval t_n by the period t_0 of the self-sustained oscillation without AC current.

3.6 Effect of taste substances on static properties of membranes

One method by which a taste sensor can be produced may be by using similar materials to biological systems as the transducer. The biological membrane is composed of proteins and lipids. Proteins are considered as the main receptors of chemical substances to produce taste, although more detailed study may be necessary to confirm this. As seen in the last section, lipid membranes may possibly be able to act as the transducer to transform information of taste to electric signals. Consequently, effects of taste substances on static (or stationary) electrical characteristics of the DOPH-adsorbed membrane were studied.

The membrane was placed between a 100 mM KCl solution and a 1 mM KCl solution to which taste substances were added. The electric potential difference across the membrane (i.e., the membrane potential) was measured by taking the origin at the 1 mM KCl solution. From comparison with the biological system, the 1 mM and 100 mM KCl sides can be regarded as the outside and the inside of the taste cell on the tongue, respectively. Changes of the membrane potential and electric resistance were studied by application of five chemical substances to produce different taste qualities.

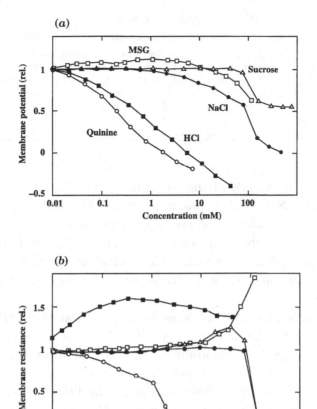

Figure 3.23. Changes in the membrane potential (*a*) and resistance (*b*) with five kinds of chemical substance to produce different tastes. The adsorbed amount of DOPH in a membrane filter was taken to be large. Consequently, the membrane resistance remained as high as several megaohms in spite of the contact with 100 mM KCl at one side because of the very slow phase transition, which takes more than one day.

The membrane potential and membrane resistance of the DOPH-adsorbed membrane are affected by taste substances. Figure 3.23 summarizes their effects.[16,53,57] The relative value is adopted by setting the initial values, typically −118 mV and 3 MΩ, to unity. The initial membrane potential changed from about −100 to −125 mV depending on the membrane preparation; in this respect, the quantitative reproducibility was not very good. This is drastically improved in a taste sensor utilizing membranes composed of lipids and polymers, as will be detailed in Chapter 6. The lower direction of the ordinate in Fig. 3.23(*a*) implies the increase of membrane potential to the positive value.

Table 3.1. *Speed of response to taste*

	Speed of response (mV/s)			
	HCl (sour)	NaCl (salty)	Quinine (bitter)	Sucrose (sweet)
DOPH system	9.6	1.7	2.2	0.03
Receptor potential	4.6	2.5	2.2	0.72

The order of magnitude of effect is quinine (bitter) > HCl (sour) > MSG (umami taste) ⩾ NaCl (salty) > sucrose (sweet). This order, as well as the values at which the change occurs, agree with those found in biological systems.[58] It is noticeable that MSG hyperpolarizes the membrane at intermediate concentrations. The membrane electric resistance is increased by MSG at any concentration while the other taste substances decrease it at higher concentrations.

There is a difference in the response to taste substances of the membrane potential and the resistance. For example, with NaCl the membrane potential changes at lower NaCl concentrations than does the resistance. This result suggests that only the surface potential is affected by the taste substances, because the resistance should also be changed if the diffusion potential within the membrane is changed. With quinine and HCl, the sign reverse of the membrane potential occurs at higher concentrations. It implies that the surface potential is increased by adsorption of positively charged quinine ions into the hydrophobic part of the membrane and binding of H^+ to the hydrophilic group of lipid molecule. Detailed analysis of these effects will be made in Section 6.3. The present results suggest one possibility for discrimination of chemical substances that show different taste qualities by using the changes in membrane potential and resistance of lipid membranes.

Table 3.1 summarizes the speed of responses to four kinds of taste substance and gives a comparison with the data obtained in biological cells.[53] The response speed is a few millivolts per second except for sweetness in the DOPH membrane system and also in biological cells. The speed of response varies in that sourness > saltiness ≃ bitterness > sweetness. This also suggests that lipid membranes would be useful as transducers to transform taste substances to electric signals.

The DOPH-adsorbed membrane responded to sweet substances. Aspartame (N-L-α-aspartyl-L-phenylalanine methyl ester), which is about 100-fold sweeter than sucrose, affected the membrane potential at a lower concentration by a factor of two.[59]

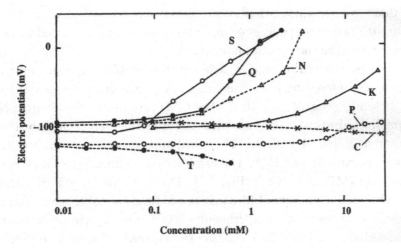

Figure 3.24. Change in the membrane potential caused by bitter substances.[52] S, strychinine; Q, quinine; N, nicotine; K, KCl; P, picric acid; C, caffeine; T, theobromine.

Let us then study effects of bitter substances on the DOPH-adsorbed membrane in more detail. Figure 3.24 shows the change in the membrane electric potential caused by bitter substances.[52] Strychnine, quinine and nicotine, which are strong bitter substances, affected the membrane potential at low concentrations of approximately 0.1 mM. Caffeine and theobromine, which are weakly bitter, scarcely changed the membrane potential.

Picric acid, which is strongly bitter, did not affect the potential. This is because of the mutual electric repulsion between anionic lipids such as DOPH and anionic species such as picric acid. This result implies that membranes made of anionic lipids such as DOPH are not sufficiently adequate in sensing chemical substances to produce taste. In the next section, the responses of a membrane composed of cationic lipids to chemical substances will be described.

As mentioned in Chapter 2, we experience modification of taste through interactions between taste-producing substances. For example, bitterness is weakened by coexistent salty or sweet substances. This suppression effect was reproduced in the DOPH-adsorbed membrane.[52] Coexistence of KCl suppressed the response of membrane potential to quinine.

A recent study using fluorometry and ESCA (electron spectroscopy for chemical analysis) has shown that quinine molecules are bound to lipid membranes by electrostatic and hydrophobic interactions.[60] The suppression effect may occur partly through a decrease in adsorption of quinine molecules onto the lipid membranes. Furthermore, it was shown that L-tryptophan, which is an amino acid with a bitter taste, is adsorbed to lipid membranes more than glycine, which tastes sweet. This result supports a hypothesis by Ney that the two taste qualities of bitterness and sweetness elicited by amino acids are related to their hydrophobicity.[61]

Let us now proceed to umami taste substances. Umami taste substances exhibit a very interesting phenomenon of taste interaction, i.e., a synergistic effect. If MSG is coexistent with IMP or GMP, a remarkable flavoring effect appears; it is called the synergistic effect of taste. The DOPH-adsorbed membrane can reproduce this effect.[16] The membrane resistance is greatly changed if MSG is coexistent with IMP. The response is amplified by a small amount of coexistent IMP, as shown in Fig. 3.25. This is a synergistic taste effect. In this condition, the quantity of MSG adsorbed to the membrane was found to increase; this agrees with an experimental result *in vivo*.[62] However, the ratio of increase of adsorbed MSG with coexisting nucleotide was considerably smaller than that found in a biological system, bovine taste papillae.

The human panel tests show that the synergistic effect can be expressed by:[63]

$$y = u + \gamma_s uv', \tag{3.20}$$

where y is the MSG concentration in the absence of nucleotide to give the equivalent taste intensity to the mixture, u the MSG concentration, v' the IMP or GMP concentration and γ_s denotes the constant to express the strength of this effect. If we assume that the magnitude of the membrane resistance change stands for the intensity of umami taste in the DOPH-adsorbed membrane, the constant γ_s can be calculated directly from eq. (3.20) using the value of the MSG concentration that caused the same resistance change for single and mixed solutions. The values $\gamma_s = 5.5 \times 10^4/M$ and $5.3 \times 10^3/M$ were obtained for 5 mM IMP and GMP, respectively. In humans,

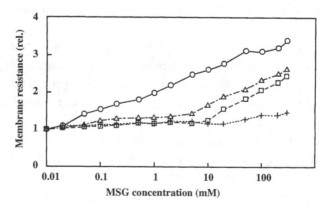

Figure 3.25. Membrane resistance change by application of MSG with coexisting IMP of 0.05 mM (+), 0.5 mM (□), 1 mM (△) and 5 mM (○).

$\gamma_s = 6.42 \times 10^4/\text{M}$ and $1.98 \times 10^5/\text{M}$ were reported.[63] The values estimated in the DOPH-adsorbed membrane are smaller than those from the human panel. In spite of this quantitative discrepancy, it is quite clear that the synergistic effect of umami taste is detectable using the lipid membrane.

Figures 3.26(*a*) and (*b*) show the surface structures of the DOPH-adsorbed membrane treated with 100 mM MSG and 1 mM IMP, respectively. Note the

(*a*)

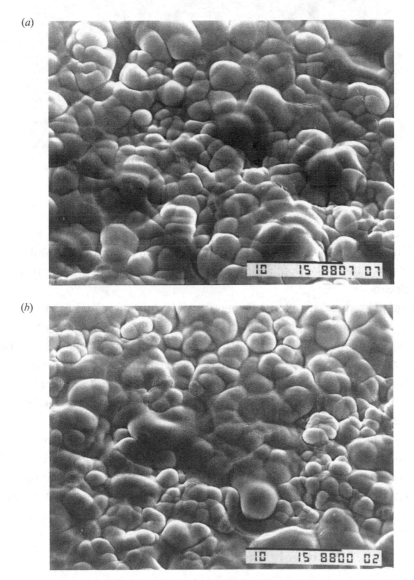

(*b*)

Figure 3.26. Surface structures of the DOPH-adsorbed membrane observed by scanning electron microscopy after treatment with (*a*) 100 mM MSG, (*b*) 1 mM IMP and (*c*) (overleaf) 1 mM IMP and 100 mM MSG.[16] The bar in the bottom right is 10 μm.

(c)

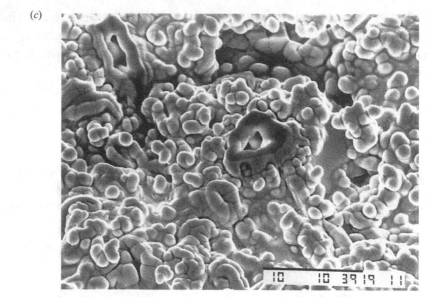

Figure 3.26 cont.

peculiar circular structures that project from the surface. Figure 3.26(c) shows the surface structure treated with a mixed solution of 1 mM IMP and 100 mM MSG. The structure is different from that of a single umami taste solution: a different structure of the lipid membrane is formed when the synergistic taste effect appears.

It seems that these structural changes caused by umami taste substances were the result of physicochemical adsorption of MSG and swelling of the oil droplets. Although addition of IMP results in only small changes in the amount of MSG adsorbed to the lipid, the structural change is drastic. Study using fluorometry showed that the adsorption of MSG onto the lipid membrane occurs by weak interaction between a phosphoric acid group of lipid molecules and an amino group of MSG.[60]

The finding of taste interactions using the electrical change in a lipid membrane is very important, because we do not intend to detect each chemical substance but rather to measure the taste itself: that is, the intention in developing a taste sensor is to measure the taste experienced by humans.

3.7 Positively charged lipid membrane

The DOPH-adsorbed membrane did not respond to picric acid, which is a strong bitter substance. We, therefore, studied a response using a different lipid membrane.[38] The synthesized lipid was dialkyldimethyl ammonium bromide $((C_{18}H_{37})_2(CH_3)_2N^+Br^-)$, which has two hydrocarbon chains and a positively charged ammonium group.

A silicon film of $350\,\mu m$ thickness with a pore of $50\,\mu m$ diameter was fabricated. Its surface was oxidized, and polymer $(C_{18}H_{37})_2(CH_3)_2N^+Br^-$ multi-bilayer complexes were cast. The polymer was sodium polystyrene sulfonate (PSS$^-$). The immobilized bilayer membranes showed electric oscillations, i.e., excitability, originating in an ordered-fluid phase transition brought about by change in ion concentration.

Figure 3.27 shows the changes in membrane electric potential when the membrane was placed between 100 mM KCl and 1 mM KCl with picric acid, acetic acid, NaCl and sucrose added to the latter. The sign of the potential is the opposite to that seen in the DOPH membranes because of the positive charge of the lipid concerned. It can be seen that picric acid is detected at relatively low concentrations. Fluctuations of electric potential or oscillations were sometimes observed on application of the picric acid; this may suggest penetration of the bitter substance into the hydrocarbon part of the lipid bilayer. By comparison, the effect of the strongly bitter quinine was weak. This tendency is in contrast to that shown in Fig. 3.24. Magnitude of change in electric potential with the bitter substance was larger at the fluid state of the membrane that occurred at higher temperatures than the ordered state at lower temperatures. These results imply that the substances first

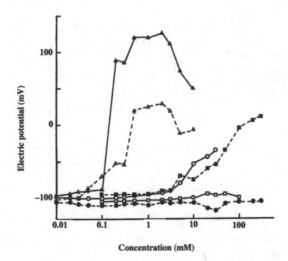

Figure 3.27. Changes in membrane potential caused by taste-producing chemical substances.[38] ▲, Picric acids at 50°C; △, picric acid at 40 °C; ■, NaCl; ○, quinine; ●, acetic acid; □, sucrose. All substances apart from picric acid were assessed only at 40 °C. The membrane was placed between 100 mM KCl and 1 mM KCl solution, to which taste substances were added. The origin of electric potential was taken at 100 mM KCl.

interact with the charge of the lipid by the electric force and then penetrate into the hydrocarbon part of lipids by the hydrophobic interaction.

The speed of response to each chemical was of the same order as the DOPH-adsorbed membrane and hence the biological system.

3.8 Summary

The last two sections have been devoted to descriptions of the effects of taste-producing chemical substances on the electrical characteristics of two kinds of lipid membrane. Five kinds of chemical substance producing different tastes are discriminated using the changes of membrane electric potential and resistance of lipid membranes. The response characteristics of the negatively charged lipid membrane are very different from those seen in the positively charged membrane. Interactions between taste substances such as the synergistic effect in umami taste can be reproduced using lipid membranes.

These results imply that lipid membranes are useful materials to transform information of taste into electric signals. Based on this finding, a taste sensor was developed; its details will be described in Chapter 6.

REFERENCES

1. Nicolis, G. and Prigogine, I. (1977). *Self-Organization in Nonequilibrium Systems.* Wiley-Interscience, New York.
2. Belousov, B. P. (1959). *Sb. Ref. Radats. Med. 1958.* Medgiz, Moscow, p. 145.
3. Zhabotinsky, A. M. and Zaiken, A. N. (1973). *J. Theor. Biol.*, 40, 45.
4. Field, R. J. and Noyes, R. M. (1974). *J. Am. Chem. Soc.*, 96, 2001.
5. Tomita, K., Ito, A. and Ohta, T. (1977). *J. Theor. Biol.*, 68, 459.
6. Jaffe, L. F. (1979). In *Membrane Transduction Mechanism*, eds. Cone, R. A. & Dowling, J. E. Raven Press, New York, p. 199.
7. Hamada, S., Ezaki, S., Hayashi, K., Toko, K. and Yamafuji, K. (1992). *Plant Physiol.*, 100, 614.
8. Borgens, R. B. (1982). *Int. Rev. Cytol.*, 76, 245.
9. Kobatake, Y., Irimajiri, A. and Matsumoto, N. (1970). *Biophys. J.*, 10, 728.
10. Yoshida, M., Kobatake, Y., Hashimoto, M. and Morita, S. (1971). *J. Membr. Biol.*, 5, 185.
11. Kamo, N., Yoshioka, T., Yoshida, M. and Sugita, T. (1973). *J. Membr. Biol.*, 12, 193.
12. Kobatake, Y. (1975). *Adv. Chem. Phys.*, 29, 319.
13. Toko, K., Nitta, J. and Yamafuji, K. (1981). *J. Phys. Soc. Jpn*, 50, 1343.
14. Binder, K. (1973). *Phys. Rev.*, 8, 3423.
15. Nitzan, A., Ortoleva, P., Deutch, J. and Ross, J. (1974). *J. Chem. Phys.*, 61, 1056.
16. Hayashi, K., Matsuki, Y., Toko, K., Murata, T., Yamafuji, Ke. and Yamafuji, K. (1989). *Sens. Mater.*, 1, 321.
17. Tien, H. T., (1974). *Bilayer Lipid Membranes (BLM), Theory and Practice.* Dekker, New York.
18. Teorell, T. (1959). *J. Gen. Physiol.*, 42, 831.
19. Kobatake, Y. (1970). *Physica*, 48, 301.
20. Monnier, A. M. (1977). *J. Membr. Sci.*, 2, 67.

21. Antonov, V. F., Petrov, V. V., Molnar, A. A., Predvoditelev, D. A. and Ivanov, A. S. (1980). *Nature*, 283, 585.
22. Dupeyrat, M. and Nakache, E. (1978). *Bioelectrochem. Bioenerg.*, 5, 134.
23. Toko, K., Yoshikawa, K., Tsukiji, M., Nosaka, M. and Yamafuji, K. (1985). *Biophys. Chem.*, 22, 151.
24. Toko, K., Ryu, K., Ezaki, S. and Yamafuji, K. (1982). *J. Phys. Soc. Jpn*, 51, 3398.
25. Toko, K., Tsukiji, M., Ezaki, S. and Yamafuji, K. (1984). *Biophys. Chem.*, 20, 39.
26. Yamafuji, K. and Toko, K. (1985). *Mem. Fac. Eng. Kyushu Univ.*, 45, 180.
27. Arisawa, J. and Furukawa, T. (1977). *J. Membr. Sci.*, 2, 303.
28. Urahama, K. and Yamafuji, K. (1982). *Trans. IECE Jpn*, 65C, 185 [in Japanese].
29. Toko, K., Tsukiji, M., Iiyama, S. and Yamafuji, K. (1986). *Biophys. Chem.*, 23, 201.
30. Iiyama, S., Toko, K. and Yamafuji, K. (1987). *Biophys. Chem.*, 28, 129.
31. Toko, K., Ozaki, N., Iiyama, S., Yamafuji, K., Matsui, Y., Yamafuji, Ke., Saito, M. and Kato, M. (1991). *Biophys. Chem.*, 41, 143.
32. Sugawara, K., Aoyama, H., Sawada, Y. and Toko, K. (1993). *J. Phys. Soc. Jpn*, 62, 1143.
33. Toko, K., Iiyama, S., Hayashi, K., Ozaki, N. and Yamafuji, K. (1988). *J. Phys. Soc. Jpn*, 57, 2864.
34. Cohen, M. H. and Turnbull, D. (1959). *J. Chem. Phys.*, 31, 1164.
35. Träuble, H., Teubner, M., Woolley, P. and Eibl, H. (1976). *Biophys. Chem.*, 4, 319.
36. Toko, K., Nakashima, N., Iiyama, S., Yamafuji, K. and Kunitake, T. (1986). *Chem. Lett.*, 1986, 1375.
37. Toko, K., Yoshida, T., Yamafuji, K., Iiyama, S., Nakashima, N. and Kunitake, T. (1986). *Proc. 6th Sensor Symp.*, p. 225.
38. Hayashi, K., Yamafuji, K., Toko, K., Ozaki, N., Yoshida, T., Iiyama, S. and Nakashima, N. (1989). *Sens. Actuators*, 16, 25.
39. Fuchikami, N., Sawashima, N., Naito, M. and Kambara, T. (1993). *Biophys. Chem.*, 46, 249.
40. Marek, M. (1984). In *Modelling of Patterns in Space and Time*, eds. Jäger, W. & Murray, J. D. Springer-Verlag, Berlin, p. 214.
41. Urahama, K. (1981). *Trans. IECE Jpn*, 64C, 453 [in Japanese].
42. Shanes, A. M. (1958). *Pharmacol. Rev.*, 10, 59.
43. Meyer, H. H. (1899). *Arch. Pharmacol. Exp. Pathol.*, 42, 109.
44. Miller, J. C. and Miller, K. W. (1975). In *MTP International Review of Science; Physiological and Pharmacological Science*, ed. Blaschko, H., University Park Press, Baltimore, MD, p. 33.
45. Ueda, I., Shieh, D. D. and Eyring, H. (1974). *Anesthesiology*, 41, 217.
46. Iiyama, S., Toko, K., Murata, T., Suezaki, Y., Kamaya, H., Ueda, I. and Yamafuji, K. (1990). *Biophys. Chem.*, 36, 149.
47. Glass, L., Guevara, M. R., Shrier, A. and Perez, R. (1983). *Physica D*, 7, 89.
48. Freeman, W. J. (1987). *Biol. Cybern.*, 56, 139.
49. Meyer, K. H. and Hemmi, H. (1953). *Biochem. Z.*, 277, 39.
50. Requena, J., Velaz, M. E., Guerrero, J. R. and Medina, J. D. (1985) *J. Membr. Biol.*, 84, 229.
51. Iiyama, S., Toko, K., Murata, T., Ichinose, H., Suezaki, Y., Kamaya, H., Ueda, I. and Yamafuji, K. (1992). *Biophys. Chem.*, 45, 91.
52. Iiyama, S., Toko, K. and Yamafuji, K. (1986). *Agric. Biol. Chem.*, 50, 2709.
53. Toko, K., Hayashi, K., Iiyama, S. and Yamafuji, K. (1987). *Dig. Tech. Papers of Transducers '87*, p. 793.
54. Hayashi, K., Toko, K. and Yamafuji, K. (1989). *Jpn J. Appl. Phys.*, 28, 1507.
55. Saida, Y., Matsuno, T., Toko, K. and Yamafuji, K. (1993). *Jpn J. Appl. Phys.*, 32, 1859.
56. Saida, Y., Matsuno, T., Toko, K. and Yamafuji, K. (1992). *Sens. Mater.*, 4, 135.
57. Iiyama, S., Toko, K. and Yamafuji, K. (1987). *Maku (Membrane)*, 12, 231 [in Japanese].

58. Pfaffmann, C. (1959). In *Handbook of Physiology*, Sect. 1 *Neurophysiology*, Vol. 1, ed. Field, J. American Physiological Society, Washington DC, p. 507.
59. Iiyama, S., Toko, K. and Yamafuji, K. (1989). *Agric. Biol. Chem.*, 53, 675.
60. Hayashi, K., Shimoda, H., Matsufuji, S. and Toko, K. (1999). *Trans. IEE Jpn*, 119-E, 374 [in Japanese].
61. Ney, K. H. (1971). *Z. Lebensm.-Unters. Forsch.*, 147, 64.
62. Torii, K. and Cagan, R. H. (1980). *Biochim. Biophys. Acta*, 627, 313.
63. Yamaguchi, S. (1967). *J. Food Sci.*, 32, 473.

4

Biosensors

4.1 Principle of biosensors

A biosensor is a kind of a chemical sensor that measures chemical substances by means of biomaterials and related materials (see Table 1.3). Biosensors can be classified by the biomaterial used into enzyme, microbial, immuno-, organelle and tissue sensors (Fig. 4.1).

A biosensor is, in principle, made by immobilizing these biomaterials and related materials to a sensing membrane and combining this with an electrochemical device. As shown in Fig. 4.2, it converts the concentration of chemical substances to be measured into light, sound and oxygen concentration by means of the biomaterial-immobilized membrane. It is possible to estimate the concentration of chemical substances by changing this quantity to an electric signal using an electrode and a thermistor. In all, a biosensor system consists of a receptor membrane where the reaction (or response) takes place, a transducer that transforms the concentration of chemical substances into an electric signal and recording equipment or computer. In an SPR (surface plasmon resonance) biosensor, the change of refractive index near the membrane surface caused by an antigen–antibody reaction is quantified by the resonance-angle change, which can be simply transformed to an electric signal using an electronic circuit.

We use an electrode as a transducer in most cases. The method to obtain electric signals is classified roughly into potentiometry and amperometry. Potentiometry is a method that detects the concentration of ions generated with the receptor membrane in the form of a change in the membrane potential in an ion-selective electrode. There are electrodes to respond to H^+, carbon dioxide, etc. Amperometry is a method to measure the electric current that flows as a result of ions produced at the membrane; ions are detected with an electrode. The oxygen electrode and hydrogen peroxide electrode are widely used.

A biosensor makes use of chemical reactions occurring in a living body, because almost all of them are selective and specific to some chemical sub-

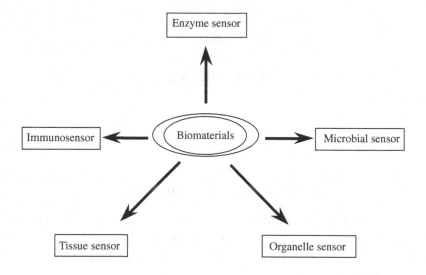

Figure 4.1. Various kinds of biosensor.

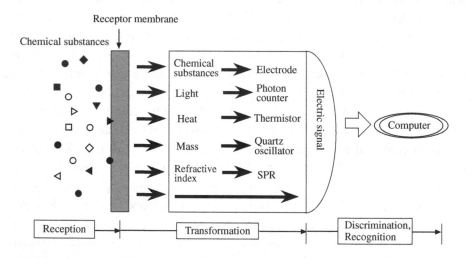

Figure 4.2. Principle of the biosensor.

stances. This high selectivity indicates that biosensors are developed using the same concepts as conventional physical sensors.

An enzyme is a catalyst that helps a chemical reaction; its interactions with its substrates resemble the relationship between a lock and its key. The enzyme is specific in which materials it will bind and it catalyzes structural changes such as addition reactions and degradation reactions. In processes such as respiration, enzymes enable a major change, such as combustion of complex molecules to give carbon dioxide and energy, to occur step by step releasing energy slowly and calmly.

In information transmission, which also involves nerve fibers, receptor proteins bind to neurotransmitters that are released into the gap (synapse) between cells. For example, in taste, the neurotransmitter is norepinephrine (noradrenaline), which is received by a specific protein in a nervous cellular membrane and causes depolarization of a membrane potential. Nerves convey information by long fibers reaching throughout the body; by comparison, hormones transmit information by being transferred all over the body through the blood vessels.

Our heart beats quickly when we take an important examination or talk in front of many people because a hormone called epinephrine (adrenaline) is released inside our blood and sends a signal indicating "I fight from now" to the whole body. The hormone released inside the blood binds to a specific protein and a reaction to send more blood from the head to the feet occurs. The hormone is related to a slow tune control of the whole body. By comparison, the nerve reports information from the outside to the brain and works in a real-time and localized fashion where it transmits information to the arm and leg, for example, in an instruction for motion as a response to a sensed stimulus.

Once we have had measles and mumps, we will not have them again or will only have a light attack. This immunity is the result of a build up of antibodies that recognize an antigen that causes sickness. An antibody binds specifically to a particular antigen and will prevent the action of the antigen. A virus cannot multiply by itself but needs a host cell in which to reproduce its genetic material. Consequently, it invades cells in order to multiply. The invaded cell and the whole body cannot function normally and illness results. Viruses contain many antigens. Figure 4.3 shows a T4 bacteriophage, which attaches to cells by its six protein feet and uses a hollow tube to inject its genetic material into the cell.

An antibody has a Y shape and is composed of four polypeptide chains in two pairs (Fig. 4.4). The short arms of the Y contain the binding sites for a specific antigen and this area varies depending on the antigen targeted. The genes coding for antibodies control the structure of different parts or domains of the protein. Varying the combinations and structures of these domain genes enables the body to produce specific antibodies to an almost infinite range of antigens from a base of approximately 500 genes in a human.

An immunosensor is a sensor that measures the concentration of an internal antigen or antibody by utilizing an antigen–antibody reaction in which this antibody is bound with an antigen.

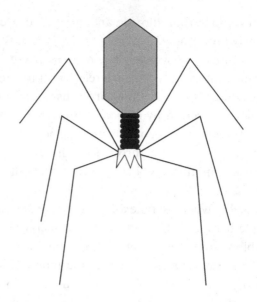

Figure 4.3. T4 bacteriophage.

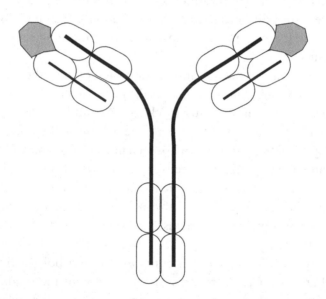

Figure 4.4. Binding of an antibody with a corresponding antigen (hatched).

4.2 Enzyme sensors

An enzyme sensor detects various chemical substances by means of enzymes with high sensitivity and selectivity. A test paper that can inspect urine easily at home is on the market. The mechanism involves impregnating the paper with an enzyme that oxidizes glucose. The reaction of the enzyme with glucose in the urine is detected using potassium iodide to give a color change with the chemical products. The higher the amount of glucose, the more active the reaction is and the deeper the color change. An enzyme sensor is made for the purpose of quantifying such a reaction with more accuracy.

It was Clark (1962) who initially proposed the principle of the enzyme sensor. The concentration of glucose can be measured indirectly by adding the membrane infiltrated with enzyme to the blood and measuring oxygen consumed by this reaction with an oxygen electrode.[1] Clark's type of oxygen electrode is composed of the anode, an oxygen-permeable polymer membrane that is in contact with an electrolytic liquid, and the cathode. Oxygen molecules permeate the polymer membrane and reach the cathode surface to be reduced electrochemically. Consequently, we can estimate the oxygen concentration by measuring the flowing electric current.

In 1966, Updike and Hicks[2] produced a glucose sensor that immobilized glucose oxidase (GOD), the enzyme that oxidizes glucose, to a membrane and measured the oxygen concentration with an electrode. By measuring the oxygen concentration, we can see the oxygen quantity that was consumed and hence the glucose concentration. Glucose is oxidized to produce hydrogen peroxide as shown in the following reaction.

$$Glucose + O_2 \rightarrow gluconic\ lactone + H_2O_2$$

By measuring hydrogen peroxide generated with this reaction, a sensor will also evaluate the glucose concentration. With the similar principle, there is also a sensor that measures saccharides such as sucrose and galactose (Table 4.1). Figure 4.5 shows the structure of the glucose sensor.

An enzyme that is included in a membrane matrix is refined and extracted usually from a microbe. The methods of immobilizing an enzyme to the membrane are classified roughly into chemical and physical methods,[3,4] as shown in Fig. 4.6. The chemical methods include formation of a covalent bond between the enzyme and a carrier formed from a macromolecule and glass (covalent bond method) and formation of covalent bond between enzyme molecules (cross-linking method) to form a non-soluble membrane. There is also an entrapment method and an adsorption method, both classified as the physical methods. An entrapment method is where the enzyme is wrapped in an insoluble macromolecule matrix, for example of collagen, polyacrylamide or silicon rubber. An adsorption method involves adsorbing the enzyme onto surface-active materials such as ion-exchange resins, carbon, clay, glass, collagen or cellulose membranes. In order to prevent the enzyme detaching, a semipermeable membrane is usually attached superficially.

Table 4.1. *Enzyme sensors for various uses*

Target	Enzyme	Immobili-zation method	Transducer	Stability (days)	Resonse time	Dynamic range (mg/l)
Glucose	Glucose oxidase	Covalent bond	O_2 electrode	100	10 s	1–500
Galactose	Galactose oxidase	Adsorption	Pt electrode	30	–	10–1000
Ethanol	Alcohol oxidase	Cross-linking	O_2 electrode	120	30 s	5–1000
Urea	Urease	Entrapment	NH_3 electrode	20	30–60 s	10–1000
Cholesterol	Cholesterol oxidase	Covalent bond	Pt electrode	80	3–5 min	0.1–100
Glutamic acid	Glutamic acid dehydrase	Adsorption	NH_3 electrode	2	1 min	10–10 000

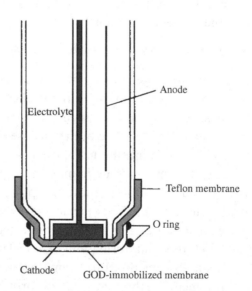

Figure 4.5. Glucose sensor using glucose oxidase (GOD).

The glucose sensor is used for diagnosis of diabetes. Since there are approximately 135 million diabetic patients in the world and there may be 300 million by the year 2025, a sensor to measure blood glucose has been strongly desired. A disposable glucose sensor (Fig. 4.7) that can be used at home to measure the glucose concentration in blood has been developed.[5] With only a drop of

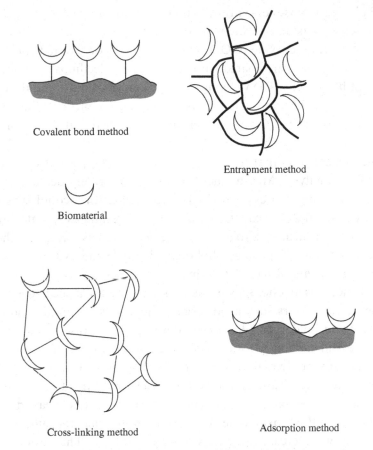

Covalent bond method

Entrapment method

Biomaterial

Cross-linking method

Adsorption method

Figure 4.6. Methods of immobilization of biomaterials in the membrane.

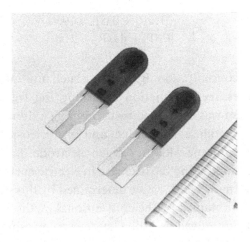

Figure 4.7. Disposable glucose sensor. (Photograph kindly offered by Dr S. Nankai in Matsushita Electric Industrial Co., Ltd, Japan.)

blood (3 µl), the glucose concentration up to 25 mM can be determined within 60 s. The sensor has an enzyme reaction layer that is a mixture of a water-soluble polymer, an enzyme (GOD) and a mediator on printed carbon electrodes. The electrodes are coated with carboxymethylcellulose to remove the influence of blood cells. The working electrode area is 1 mm^2 and the counter electrode area is 7.6 mm^2. The temperature dependence of the response current is less than 2%/°C and the sensor can be stored at 40 °C for over 3 months with no change in the initial performance.

Another sensor, one that measures galactose, a component of lactose, is used for food analysis. Measurement of alcohol is widely demanded in the food manufacturing process and the medical field. An alcohol sensor uses alcohol oxidase (alcohol oxidation enzyme) and estimates the alcohol concentration by measuring hydrogen peroxide generated as a result of the reaction. Alcohol oxidase changes alcohol to an aldehyde and hydrogen peroxide. However, this enzyme also oxidizes ethyl alcohol, methyl alcohol, n-propanol, etc., and hence this alcohol sensor has no selectivity for a specific alcohol.

Furthermore, sensors such as an organic acid sensor, amino acid sensor, urea sensor and lipid sensor have been developed. Because of the operative convenience and suitability for analysis of small quantities of samples, these sensors have great potential for use in medical applications.

As one of these applications, a sensor has been developed to evaluate freshness of fish (i.e., K value[6]) using three different sensors measuring IMP, inosine (HxR) and hypoxanthine (Hx).[7] It is known that the sum of inosine and hypoxanthine increases with the storage time of fish. The freshness index, called the K_I value, is proposed as follows.

$$K_I = \frac{([HxR] + [Hx]) \times 100}{[IMP] + [HxR] + [Hx]}, \tag{4.1}$$

where [ ] means the concentration. By utilizing such multiple selective sensors, we can obtain necessary information regarding the freshness of fish. An enzyme sensor system was developed by combining a double membrane consisting of a 5′-nucleotidase membrane and a nucleoside phosphorylase–xanthine oxidase membrane with an oxygen electrode. Each nucleotide concentration was measured using the decrease in electric current. One assay was completed within 20 min. The K_I value determined by this method had a good agreement with that obtained by the conventional method.

This method was improved as a flow system comprising the oxygen electrode and an enzyme column.[8] As a result, the lifetime of the enzyme increased and maintenance became easier.

Histamine increases with the storage time of fish and causes food poisoning. Its quantity can be measured using a colorimetric analysis at 510 μm after pretreatment using an ion-exchange column. Recently, a simpler method to measure histamine without the pretreatment was proposed by Ohashi *et al.*, where the oxygen electrode and purified amine oxidase were incorporated into the commercial apparatus to measure the K_I value.[9]

The lipid bilayer is the basic structure of all biological membranes. Proteins are embodied in this structure to work as an ion channel, pump or receptor. It would appear, therefore, that a lipid bilayer can provide optimal conditions for the functioning of such proteins. Based on this idea, biosensors using supported bilayer lipid membranes (sBLM) that incorporate enzymes have been developed.[10–12]

The sBLM is deposited on a metallic support so as to face the aqueous phase on the other side. Applications of the sBLM to biomolecular electronic devices and biosensors are expected. Ammonia concentration was measured using a biosensor that utilizes avidin-modified urease bound to sBLM, which is supported on a conducting polymer, polypyrrole, adjacent to the metal surface.[10] The detection limit was 5–10 mM urea. This value is larger than that for the urea FET biosensor (~1 mM, see below), for example. It is considered that the sensitivity will be improved by choice of an adequate metal support.

4.3 Microbial sensors

A microbial sensor is made to detect chemical substances utilizing respiratory function and metabolism of a microbe; it is used in the fermentative industrial processes and in environment measurements.

This sensor is classified into two types. The first one is the sensor that measures a change of a respiratory activity of the immobilized microbe with an electrochemical device, and the other is the sensor that measures materials produced by the microbe that easily react with the electrode. Aerobic microbes use oxygen and produce energy. By measuring the oxygen consumed by means of an oxygen electrode, we can evaluate the respiratory activity. Oxygen permeates a Teflon membrane and is reduced on a platinum electrode. If a material could influence respiratory activity inside a test solution, its concentration could be evaluated from the oxygen concentration measurement. The principle, as shown in Fig. 4.8,[13] is quite similar to that of the enzyme sensor.

By immobilizing a microbe called *Pseudomonas fluorescens*, a glucose sensor is also developed.[14] When this microbe is immobilized inside a collagen

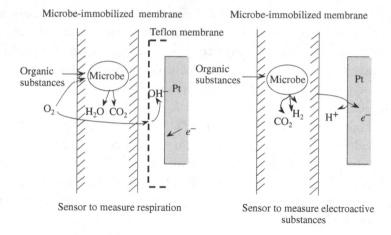

Figure 4.8. Two types of microbial sensor.[13]

membrane and the sensor is immersed in a test solution, the respiratory activity is high through utilization of glucose; hence, the quantity of oxygen spreading to the electrode decreases. As a proportional relation holds between the observed electric current (of the order of several tens of microamperes) and the glucose concentration measured by this change in respiratory activity, the glucose concentration can be evaluated easily. The threshold concentration of glucose that can be detected is 2 mg/l. The electric current was reproducible within ±6% of the relative error for 10 mg/l glucose. The durability was over two weeks.

The biological oxygen demand (BOD) is a measure of the organic matter (pollution) in water and is used to assess the quality of water. These contaminants are decomposed by microbes in a process that consumes oxygen; hence the extent of contamination can be determined by measuring oxygen. A sensor is made by immobilizing *Trichosporon cutaneum* of yeast in a cellulose membrane and fixing the membrane to an oxygen electrode.[15] The electric current is proportional to the BOD of waste water. Relative error of the BOD estimation was within ±10%. Conventional measurements of BOD would take five days; this method could produce a result in 30 min.

4.4 Integrated type biosensors

The concentration of ions generated with the membrane of a biosensor can be transformed into an electric signal by means of various electrodes. Miniaturization of the sensor using an ion-sensitive field-effect transistor (ISFET), semiconductor instead of the electrode, has been studied recently.

ISFET was proposed by Bergveld in 1970.[16] The gate part is covered by an insulation layer and immersed in an electrolytic liquid. Because the drain electric current changes with the ion concentration in the test solution, the ion concentration can be measured. The drain electric current flowing between the drain and source can be controlled with a small voltage change of the gate division. Consequently, the very small change of ion concentration is amplified to become a large change of drain electric current.

An H^+-sensitive sensor with a silicon nitride (Si_3N_4) layer as the insulation layer was developed to increase the sensitivity and stabilization of the method.[17] However, the selectivity to H^+ of this ISFET is low; it also responds to Na^+. Consequently, ISFET using a tantalum oxide (Ta_2O_5) as the sensitive membrane was developed.[18]

If we attach the enzyme-immobilized membrane to a gate division of ISFET, a transistor to respond to a specific ion, the concentration of chemical substances can be measured by detecting ions released from the immobilized membrane. This sensor device is called an enzyme FET (ENFET). A sensor can be miniaturized using a transistor and integration of various sensors may be possible. It may be possible in future to implant such a sensor in the body to assist in management of good health.

In 1980, Caras and Janata[19] announced a FET that is able to measure penicillin concentration. The reaction is:

$$\text{Penicillin} + H_2O \rightarrow \text{penicilloic acid}$$

Penicillinase acts as a catalyst for this reaction. If we immobilize penicillinase on a membrane and immerse this in the test solution, H^+ is produced because penicilloic acid is a strong acid. The pH change is measured with an H^+-sensitive FET. Together with FET as a reference electrode to give a standard for the electric potential, the enzyme FET was integrated on a silicon board.

A sensor that measures urea is based on the decomposition of urea by urease (urea lytic enzyme):[20]

$$\text{Urea} + 2H_2O + H^+ \rightarrow 2NH_4^+ + HCO_3^-$$

As urea is utilized, NH_4^+ is generated and the pH increases. Consequently, the urea concentration can be estimated by measuring the change in pH with ISFET. The enzyme FET for urea is illustrated in Fig. 4.9.

Some technical expertise is necessary to construct the sensor, for example in order to glue the membrane onto the gate. In this system, silane treatment was used to prevent the membrane stripping from the gate. The gate surface of the ISFET is Si_3N_4 treated with γ-aminopropyl triethoxylane (γ-APTES). The ethoxy group of γ-APTES is bound to the surface of Si_3N_4, and the amino

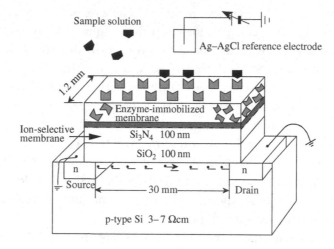

Figure 4.9. Urea biosensor of FET type.

group is superficially exposed. The urease immobilized membrane treated with glutaric aldehyde is not stripped because of chemical bond formation between the amino group and glutaric aldehyde.

The urea FET can be used in the range 0.05–10 mg/ml. The output voltage is shown in Fig. 4.10 and changes linearly from several millivolts to approximately 50 mV. It is possible to obtain a good response using the FET once per day for about one month. The urea concentration is 0.1–0.2 mg/ml in the blood of a normal person, which is within the effective range of the urea FET.

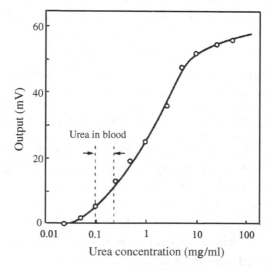

Figure 4.10. Output of the urea FET biosensor.[20]

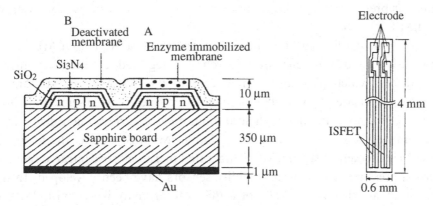

Figure 4.11. One-chip biosensor.[21]

Figure 4.11 is a one-chip biosensor composed of one pair of ISFET and a gold electrode as a pseudo reference electrode on one chip.[21] Generally the electric potential of gold is unstable in aqueous solution and it does not give a reliable value. In Fig. 4.11, urease is immobilized at one side (A) of one pair of ISFET, with the membrane with no enzyme activity on the other side (B). By taking an output difference from the two FET, we can cancel the electric potential generated by the gold electrode as a common standard. The pH change that occurs inside the urease-immobilized membrane is picked up with FET (A), and the pH in the test solution is measured with FET (B). A sapphire basis board was established in order to isolate the FET electrically from the gold electrode. The enzyme-immobilized membrane is made by photolithography.

Figure 4.12 is an integrated type amino acid sensor to measure four amino acids (L-lysine, L-glutamic acid, L-arginine, L-histidine).[22,23] Measurement of amino acids is frequently required in many fields such as medicine and food industries (for quality control), and this sensor selectively responds to amino

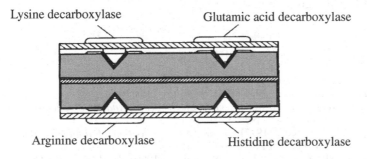

Figure 4.12. Integrated biosensor to measure four amino acids.[23]

acids. At present, however, the measurement range is narrow, for example 1–3 mM for lysine.

Development of integrated microelectromechanical systems (iMEMS) is vigorously pursued in current research.[24] Integrated circuit (IC) micro-transducers, which play an essential role in the iMEMS, are sensors or actuators fabricated by CMOS (complementary metal-oxide semiconductor) or IC technology combined with additional bulk or surface micromachining, thin-film deposition or electroplating. Some iMEMS such as the magnetic angle measurement system, CMOS integrated thermoelectric infrared sensors, thermal pressure sensors and humidity sensors have been developed. In the field of chemical sensors, study of a micro total analysis system (μTAS) is growing rapidly.[25-27] Manz *et al.* proposed a high-pressure liquid chromato-graphic method that involves a silicon chip with an open-tubular column (6 μm × 2 μm × 15 cm) and a conductometric detector using Pt as an electrode, a chip holder and an injector. The μTAS is applied to many chemical analysis systems such as DNA analysis, mass analysis, adsorptiometric flowcell and capillary electrophoresis.

The on-line μTAS was developed for the continuous measurement of extra-cellular glutamate.[28] Glutamate plays an important role in synaptic efficacy, including the long-term potentiation and long-term depression of excitatory neurotransmission. Therefore, the continuous monitoring of neurotransmit-ters is an essential technique for studying the physiology of nerve cells. The sensor developed using a micromachining technique is constructed from two glass plates, one of which contains a rectangular flow channel connected to sampling and outlet capillaries, the other being a carbon film-based electro-chemical cell. An enzyme (glutamate oxidase)-modified electrode surrounded by a polymer film is formed on the electrochemical cell together with reference and counter electrodes. The increase in extracellular glutamate caused by stimulation of a nerve cell using γ-aminobutyric acid (GABA) was success-fully monitored using this biosensor. A similar system is applied to measure-ments of lactate, glucose and acetylcholine.

REFERENCES

1. Clark, L. C. and Lyons, C. (1962). *Ann. N. Y. Acad. Sci.*, 102, 29.
2. Updike, S. J. and Hicks, G. P. (1967). *Nature*, 214, 986.
3. Kobos, R. K. (1980). In *Ion-Selective Electrodes in Analytical Chemistry*, Vol. 2, ed. Freiser, H. Plenum Press, New York, p. 1.
4. Chang, T. M. S. (ed.) (1977). *Biomedical Application of Immobilized Enzymes and Proteins*, Vol. 1, Plenum Press, New York.
5. Kawaguri, M., Yoshioka, T. and Nankai, S. (1990). *Denki Kagaku*, 58, 1119.
6. Saito, T., Arai, K. and Matsuyoshi, M. (1959). *Bull. Jap. Soc. Sci. Fish*, 24, 749.

7. Karube, I., Matsuoka, H., Suzaki, S., Watanabe, E. and Toyama, K. (1984). *J. Agric. Food Chem.*, 32, 314.
8. Okuma, H., Takahashi, H., Yazawa, S., Sekimukai, S. and Watanabe, E. (1992). *Anal. Chim. Acta*, 260, 93.
9. Ohashi, M., Nomura, F., Suzuki, M., Otsuka, M., Adachi, O. and Arakawa, N. (1994). *J. Food Sci.*, 59, 519.
10. Hianik, T., Červeňanská, Z., Krawczynski vel Krawczyk, T. and Šnejdárková, M. (1998). *Mater. Sci. Eng.*, C5, 301.
11. Tvarožek, V., Tien, H. T., Novotný, I., Hianik, T., Dlugopolský, J., Ziegler, W., Leitmannová-Ottová, A., Jakabovič, J., Řeháček, V. and Uhlár, M. (1994). *Sens. Actuators*, B18-19, 597.
12. Gizeli, E., Lowe, C. R., Liley, M. and Vogel, H. (1996). *Sens. Actuators*, B34, 295.
13. Seiyama, T., Shiokawa, J., Suzuki, S. and Fueki, K. (1982). *Chemical Sensor*. Kodansha, Tokyo, Ch. 4 [in Japanese].
14. Karube, I., Mitsuda, S. and Suzuki, S. (1979). *Eur. J. Appl. Microbiol. Biotechnol.*, 7, 343.
15. Karube, I., Matsunaga, T., Mitsuda, S. and Suzuki, S. (1977). *Biotechnol. Bioeng.*, 19, 1535.
16. Bergveld, P. (1970). *IEEE Trans. Biomed. Eng.*, BME-17, 70.
17. Matsuo, T. and Wise, K. D. (1974). *IEEE Trans. Biomed. Eng.*, BME-21, 485.
18. Matsuo, T. and Esashi, M. (1981). *Sens. Actuators*, 1, 77.
19. Caras, S. and Janata, J. (1980). *Anal. Chem.*, 52, 1935.
20. Miyahara, Y., Moriizumi, T., Shiokawa, S., Matsuoka, H., Karube, I. and Suzuki, S. (1983). *Nihon Kagaku-kai Shi*, 823 [in Japanese].
21. Kuriyama, T., Kimura, J. and Kawana, Y. (1985). *NEC Research & Development*, 78, 1.
22. Suzuki, H., Tamiya, E. and Karube, I. (1990). *Anal. Chim. Acta*, 229, 197.
23. Karube, I. and Tamiya, E. (1994). *Bioelectronics*. Asakura Shoten, Tokyo, p. 121 [in Japanese].
24. Baltes, H. (1996). *Sens. Actuators*, A56, 179.
25. Terry, S. C., Jerman, J. H. and Angell, J. B. (1979). *IEEE Trans. Electron Devices*, ED-26, 1880.
26. Manz, A., Miyahara, Y., Miura, J., Watanabe, Y., Miyagi, H. and Sato, K. (1990). *Sens. Actuators*, B1, 249.
27. Woolley, A. T., Hadley, D., Landre, P., de Mello, A. J., Mathies, R. A. and Northrup, M. A. (1996). *Anal. Chem.*, 68, 4081.
28. Niwa, O., Horiuchi, T., Kurita, R., Tabei, H. and Torimitsu, K. (1998). *Anal. Sci.*, 14, 947.

5

Odor sensors

5.1 Types of odor sensor

An odor-sensing system is required in various fields such as the food, drink, cosmetics and environmental industries. Human sensory evaluations are often affected by physical and mental conditions. Consequently, development of an odor-sensing system to provide an objective, quantitative estimate is necessary.

Detailed molecular components in gas can be analyzed using conventional gas chromatography. However, it must be emphasized that even if we detect the molecular components, we cannot identify the odor directly. Moreover, real-time measurement is not easy as this method usually takes a few hours.

The sense of smell involves receiving chemical substances, as in the case of taste. The noticeable property is the diversity of the sensitivity and selectivity. Dogs have much greater sensitivity than humans by a factor of 10^6–10^8, as shown in Table 5.1. Insects have extremely high selectivity and sensitivity to one particular molecule, known as a pheromone, which binds to one type of olfactory cell. Experts in the cosmetics industries can also detect one kind of flavor (presumably originating from one type or several types of molecule) among several thousands of mixed flavors. In our daily life, however, we comprehensively express the quality of odor of foodstuffs, which originates from many kinds of chemical molecule, when we take dinner, for example. The sense of smell has two opposite properties. One is the recognition of odor without separating the gas producing the odor into its single chemical components. The other property is that one or a few molecules such as pheromone are detected by one olfactory cell on average with extremely high sensitivity and selectivity. At present, we have no answer to quite a simple question about whether the reception mechanism is the same for these two properties or not. Of course, odor identification is different from gas identification in most cases, even if the gas comprises plural kinds of chemical substance, because the odor is identified by the reception of the gas and then the perception by living systems.

Table 5.1. *Thresholds for odor substances in dogs and humans*

Odor substances	Molecules per ml water	
	Dogs	Humans
Acetic acid	5.0×10^5	5.0×10^{13}
Propionic acid	2.5×10^5	4.2×10^{11}
Valeric acid	3.5×10^4	6.0×10^{10}
Butyric acid	9.0×10^3	7.0×10^9
Caproic acid	4.0×10^4	2.0×10^{11}
Caprylic acid	4.5×10^4	2.0×10^{11}

Persaud and Dodd proposed the possibility of an electronic nose with broadly tuned receptors in 1982.[1] Responses to odorants are obtained not only in olfactory systems but also in nonolfactory systems, such as frog taste cells and the neuroblastoma cells, which contain no specific receptor proteins for odorants.[2,3] It seems that almost all the research effort is based on the mechanism to treat output signals from many receptors that have no specificity to an odorant molecule.

Let us explain briefly the situation where the odor is perceived without decomposing it into the constituents, with the aid of a slightly modified model that was used to explain the feasibility of multivariate analysis.[4] The abscissa in Fig. 5.1(*a*) is the axis for discriminating molecules (e.g., molecular weight, retention time in the case of gas chromatography), whereas the ordinate indicates the concentration of a molecule. $M_1, M_2, M_3, \ldots$, are the detected molecules. Even if these molecules are detected, the odor produced by the gas cannot be reproduced because the contribution of each molecule to the receptor cell of the human olfactory system is different and we have no knowledge about it. Therefore, we must construct new axes instead of those in Fig. 5.1(*a*). These are the abscissa composed of groups of molecules to show the similar odor and the ordinate to express the strength felt by humans.

Of course, realization of odor sensor to reproduce Fig. 5.1(*b*) is impossible at present, mainly because we have no idea of the reception mechanism in olfactory systems. Nevertheless, most odor sensors developed recently seem to have one striking property, which is to use several nonspecific receptors. Output patterns from these receptors are treated using multivariate analysis or neural networks for odor identification. This situation is similar to biological systems although sensing characteristics may be different, as shown in Fig. 5.1(*c*), which differs from Fig. 5.1(*b*) in the sensitivity to each molecule.

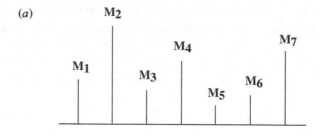

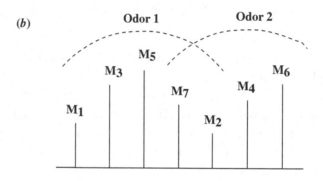

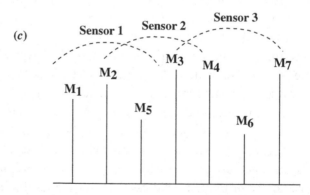

Figure 5.1. Detection of molecules and odor identification (see text for details).

In a practical sense too, this method of arranging plural sensors is effective.[4] The sensitivity of a gas sensor is generally low. Hence, even if a gas molecule M_1 is an objective gas to be detected, another kind of gas molecule (M_2) is also detected by the sensor S_1. To subtract the signal M_2 from S_1, another sensor such as S_2 is necessary. However, it may detect M_3 and M_4, too. It implies that a third sensor, S_3, is required. Gas (and sometimes odor) identification might be possible by an array of gas sensors, even if the sensors have low selectivity.

For high-sensitive detection of one molecule such as a pheromone, the situation may be very different. In this case, the above method does not seem to be effective; hence, a novel sensor with high selectivity to one molecule must be developed. Whereas a different concept of design and mechanism is necessary for this purpose, a promising way may be to use a receptor molecule in an artificial odor-sensing system by extracting it from the biological membrane. This chapter is devoted to the description of odor sensors using plural sensors with low selectivity.

Developed odor sensors are briefly classified into two types based on the materials involved: biomimetic and nonbiomimetic. Odorant molecules are relatively hydrophobic and hence are adsorbed to the hydrophobic portion of a bilayer lipid membrane in an olfactory receptor. Output patterns from many receptors with slightly different properties are processed by an olfactory neuron network to identify the odor, as above. Therefore, some researches utilize lipid membranes as a material to receive odorants.

5.2 Odor sensor using a quartz oscillator coated with lipid membranes

Rock crystal has an interesting property in that it is distorted by application of electric voltage while an electric field is generated by application of pressure. This effect is called the piezoelectric effect (see also Table 1.2). Because of this effect, the rock crystal shows a stable self-sustained oscillation of the electric voltage across it when an AC voltage, the frequency of which is near the resonance frequency, is applied using an external oscillatory circuit.

The resonance frequency is changed with the mass of the quartz crystal. Let ΔM and Δf denote the changes in the mass and the frequency, respectively, then Δf is given by

$$\Delta f = -cf^2 \frac{\Delta f}{A}, \tag{5.1}$$

where c is a positive constant, f the resonance frequency (about 10 MHz for the AT-cut quartz resonator) and A is the electrode area. Figure 5.2 shows the structure of the odor sensor using the quartz crystal microbalance (QCM).[5,6]

The principle of this sensor is to detect the frequency shift Δf brought about by the mass change ΔM associated with adsorption of odorant molecules to the sensing membrane on the quartz. If plural sensors with different sensing membranes are used and show different frequency changes for an odorant gas, the odor can be identified by analyzing their output pattern.

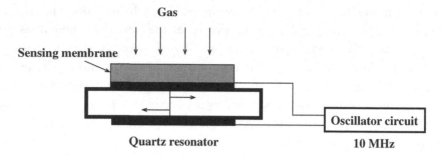

Figure 5.2. Odor sensor utilizing a quartz resonator.

Several different types of polymer and lipid multi-bilayer were used as the sensing membrane materials. The output pattern was treated with a neural-network algorithm. In principle, the frequency shift becomes 0.37 Hz for the change of 1 ng. So, the adsorption of 10 ng odorants causes a shift of several hertzs in the frequency, which can be detectable.

A relationship between the threshold of odorants detected by humans and the partition coefficient of odorants was studied using a membrane composed of a lipid (dialkyldimethyl ammonium bromide, $(C_{18}H_{37})_2(CH_3)_2N^+Br^-$) and a polymer (sodium polystyrensulfonate, PSS^-).[7] The partition coefficient was calculated by dividing the adsorption amount in the sensing membrane, which is estimated from Δf, by the concentration of odorants in the vessel. Seven different odor substances, β-ionone, coumarin, citral, 1-octanol, isopentyl acetate, methyl acetate and diethyl ether, were used in the experiment. It was found that the odor sensor utilizing a quartz resonator coated with lipid multi-bilayers has a high correlation with the olfactory sensitivity of humans. This result implies that odorant substances investigated here, which are more strongly adsorbed to the lipid multi-bilayers, can also be detected at lower concentrations by humans. This result is interesting from the viewpoint of the reception mechanism at the olfactory cell.

The system is composed of several quartz sensors with different lipid or polymer materials.[5] Figure 5.3 shows the measurement system. The adopted membrane materials were (a) epoxy resin, (b) triolein, (c) squalane, (d) acetyl-cellulose plus triolein, (e) acetylcellulose plus diethylene glycol and (f) di-*n*-octyl phthalate. Acetylcellulose was chosen to increase the adsorption surface of the resonator. Squalane, diethylene glycol and di-*n*-octyl phthalate are the materials used in gas chromatography and have different gas adsorption specificity. Each material was dissolved in acetone and coated on the electrode surface of a quartz resonator by dipping the resonator into the solution. The thickness of the coated membrane was estimated at 1 μm or less using the frequency shift.

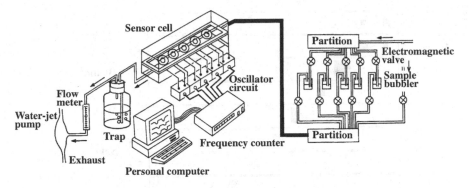

Figure 5.3. Measurement system of odors.[5]

Six quartz resonators with different membranes were installed in a gas-flow cell and connected to CMOS oscillator circuits. The frequency shifts of all the resonators were measured using a six-channel frequency counter and input to a computer. Two gas-flow routes were prepared. One is for the flow of dried air, and the other is for the flow of sample gas from a sample solution in a bubbler. They were switched using electromagnetic valves. The volume of the sensor cell was 1.5 ml and the flow rate was about 100 ml/min.

A three-layer neural network was prepared and the back-propagation algorithm was used in recognition of odor. The result of a trial to recognize 11 kinds of alcoholic drink is shown in Fig. 5.4.[5] The output units from 1 to 11 correspond to the 11 drinks. The areas of the squares for these output units are proportional to the magnitudes of outputs. If the large squares are aligned on the diagonal, the discrimination of samples is successful. Learning of the neural network was made 18 000 times with a set of 14 measurements of the above 11 samples. The input data for training was again input to the neural network, and the outputs were studied. It can be seen that the discrimination is fairly good.

In order to enhance the odor-recognition ability, the data vectors for the alcoholic drinks were put into the network after subtracting the data patterns of pure ethanol solution from those of the drinks. As a result, the recognition probability was considerably improved to 81%.

In this way, the discrimination made by ordinary persons can be copied using the odor sensor with a quartz-resonator array. Further improvements have been made in selection of membrane materials, a method to supply odor gas, temperature control, rinsing of the flow system and the recognition algorithm. It results in a successful discrimination of five brands of whiskey originating from the same company, which is not easy for the average

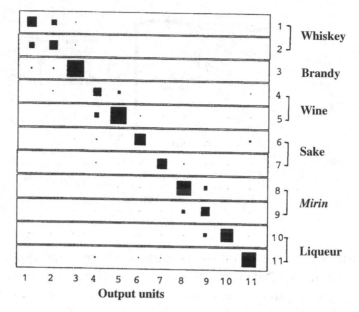

Figure 5.4. Use of an odor sensor to identify 11 alcoholic drinks.
Activity levels of output units. *Mirin* is a Japanese cooking sake.[5]

individual. In this case,[8] different lipid materials, dioleyl phosphatidylserine, sphingomyelin, lecithin and cholesterol, were adopted.

Odor discrimination was also attempted for a mixed odor containing a foreign odor resulting from various small amounts of another substance. Figure 5.5 shows the result of PCA applied to the sensor outputs for five flavor samples, 0, 1, 2, 3 and 4, implying the orange flavor plus 0, 0.1, 0.2,

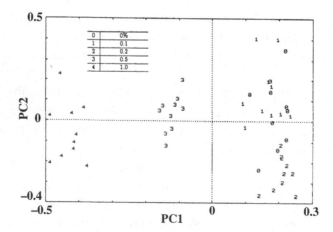

Figure 5.5. Discrimination of mixed gas flavor using the odor sensor with quartz-resonator array.[9] All the data were normalized.

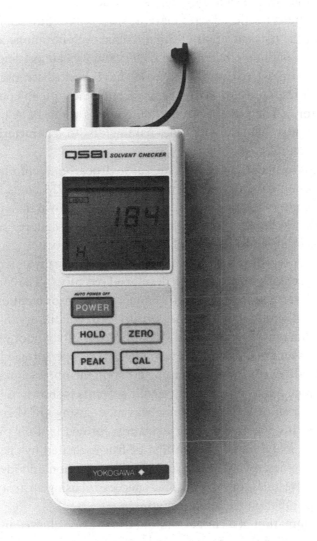

Figure 5.6. Commercial handy checker for organic solvents. (Kindly provided by
Dr G. Matsuno of Yokogawa Electric Corp., Japan.)

0.5 and 1.0%(v/v) 2-butylidene cyclohexanone, respectively.[9] Samples containing 0.5 and 1.0%(v/v) foreign substances are discriminated well from the samples containing 0, 0.1 and 0.2%(v/v). This discrimination ability reaches almost the same level as in humans. The sensing membrane with the highest contribution to the discrimination was ethylcellulose. The relationship between the concentration of 2-butylidene cyclohexanone and the output of the sensor with an ethylcellulose membrane was almost linear.

Three wines (red, white and rose) have been discriminated by utilizing transient response curves for aromas from wine with an epoxy-coated quartz resonator.[10] The pattern-recognition analysis using PCA or neural-network analysis was made using nine parameters to characterize the transient response curves.

Volatile organic compounds such as hydrocarbons, chlorinated compounds and alcohols were detected using an array of QCM coated with side-chain-modified polysiloxanes.[11] In this system, it was shown that an artificial neural network was more effective than a partial least-squares regression for the analysis of mixtures of more than two components.

An odor sensor using an array of QCM coated with a mixture of lipids and polyvinyl chloride has also been developed recently.[12] The coated materials are the same as those used in a multichannel taste sensor mentioned in Chapter 6. High correlations of the response sensitivity to volatile organic substances with that of humans were obtained. Identification of six odorants (amyl acetate, ethanol, acetic acid, water, citral and nerol) was performed by comparing the output patterns of five sensors using different membrane lipid compositions. A commercial handy checker for organic solvents is shown in Fig. 5.6.

Assessment of fish storage time was made using QCM coated with metalloporphyrins, which are cyclic structures formed by four pyrrole rings complexed by metal ions.[13] The aroma of fish is elicited by chemical substances such as amines, alcohols and sulfides. Classification and identification of the degree of freshness of stored fish was successful using an array of QCM coated with four different metalloporphyrins.

Discrimination of a chiral compound was also achieved using QCM coated with a chiral polymeric receptor.[14] The enantiomers of N-trifluoroacetyl-alanine methyl ester are preferentially bound to specific enantioselective (R)- and (S)-octyl-Chirasil-Val polymers.

5.3 Metal oxide gas sensors

Let us mention here gas identification using an array of metal oxide gas sensors. Among metal oxides, tin oxide is an intrinsically n-type bulk semiconductor. The mechanism from which the gas sensitivity originates can be understood from the following reaction:[15]

$$n + \tfrac{1}{2}O_2 \leftrightarrow O(s)^-, \qquad (5.2a)$$

$$R(g) + O(s)^- \rightarrow RO(g) + n, \qquad (5.2b)$$

where n is an electron from the conduction band of the semiconductor, the symbols s and g implying surface and gas, respectively. Equation (5.2a) implies that oxygen is physicochemically adsorbed onto lattice vacancies in the semiconductor; thus, the conductivity becomes lower. However, an electron is generated by the reaction with a combustible gas R(g) through eq. (5.2b). It leads to the increase in conductivity by the gas. The increasing sensitivity is realized by incorporation of a small amount of catalytic metals.

Since water adsorbs strongly to the surface of tin oxide and changes the response characteristics, the sensors are usually operated between 300 and 500 °C.

For discrimination of different coffee blends and roasts, an array of 12 metal oxide gas sensors was used.[16,17] TGS 800 is designed for use in ventilation control equipment and detects carbon monoxide, hydrogen, cigarette smoke and alcohol vapors; TGS 815 is used for general purposes (methane, combustible gases); TGS 821 is designed for hydrogen; TGS 825 is designed for hydrogen sulfide; and TGS 882 is designed for alcoholic vapor. These did discriminate the various coffees, as shown below.

Let i and j denote the sensor and odor, respectively, then the conductance change ΔG_{ij} of sensor i for the odor j is normalized as follows:

$$x_{ij} = \Delta G_{ij}/G_{ij}^{air}, \qquad (5.3)$$

where G_{ij}^{air} implies the steady-state conductance in air. The vector constructed from x_{ij} was utilized in the pattern recognition algorithm.

Figure 5.7 shows the result of discrimination of (a) medium roast of blend 1, (b) dark roast of blend 1 and (c) dark roast of blend 2.[17] A success rate of classification was 89.9% using the vector of eq. (5.3). It increased to 95.5% when the vector was normalized.

As tin oxide gas sensors are particularly sensitive to combustible substances, change in the odor of coffee with roasting time can be detected. Eleven sensors among the 12 sensors used showed the increasing responses with roasting time as expected.[17]

Discrimination between trimethylamine (TMA) and dimethylamine (DMA) was carried out using zinc oxide thin-film gas sensors.[18] An aluminum-doped zinc oxide sensor exhibited a high sensitivity and an excellent

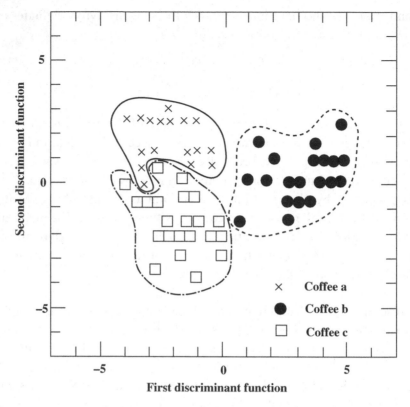

Figure 5.7. Discrimination of different coffee blends and roasts by
tin oxide gas sensor array.[17]

selectivity for amine gases. A neural network was used to make pattern recognition using parameters that characterize the transient responses of the sensors for exposure to gases. The recognition probability was 90% for TMA and DMA with the same concentration, while it reached 100% for the different concentrations.

Discriminations of wines of the same type but produced in different vineyards was tried using an array of tin oxide gas sensors.[19] The differences among these wines result from the soil characteristics and agronomic features of the vineyard, as the type of original grape was the same. This study indicated that the array of gas sensors was superior to the standard chemical analysis.

5.4 Odor sensors using conducting polymers

Organic conducting polymers derived from aromatic or heteroaromatic compounds have been used as gas and odor sensors.[20] Poly(pyrrole) was first prepared electrochemically in 1968[21] and has been most extensively studied.

The films of poly(pyrrole) consist of polycationic chains of pyrroles, the positive charges of which are counterbalanced by anions from the electrolyte solution. Pyrrole can also be polymerized inside polymers such as polyvinyl chloride.

Conducting polymers show reversible changes in conductance when chemical substances (e.g., methanol, ethanol, ethyl acetate) adsorb and desorb from the film surface. The mechanism by which the conductance is changed by adsorption of chemical molecules is not clear at present. Conducting polymers do not have high specificity for each gas. The dependence of response on the concentration of, for example, methanol is almost linear over a wide range from 0 to 15 mg/l.

A sensor array was constructed from 20 different polymers made from modified monomer units, which show broad overlapping response patterns for different volatile compounds.[20] The sensor responses were normalized to produce the concentration-independent patterns. Figure 5.8 shows the

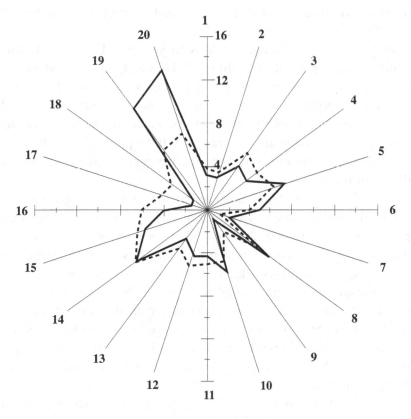

Figure 5.8. Response patterns (normalized data) of an array of conducting polymer sensors for two perfumes, musk (——) and ylang ylang (- - - - -).[20]

response patterns for two different perfumes. Whereas all the sensors respond to some extent, some groups of sensors show large responses. The response patterns for two perfumes are clearly distinguishable. Application of the neural network algorithm to the response patterns can make it possible to assess the odor quality.

Discrimination of wine flavors using conducting polymers will be discussed in Chapter 8.

5.5 Gas-sensitive field-effect devices

Lundström *et al.* proposed identification of molecules using gas-sensitive field-effect devices,[22–24] where the response is the shift of the $I_D(V_G)$ curve along the voltage axis. The change of the gate voltage V_G is measured to compensate for the change in I_D caused by gas molecules. Eleven sensors were used to measure methanol, ethanol, 1-propanol and 2-propanol. Three metal oxide gas sensors were included in the 11 sensors, as detailed in the caption of Fig. 5.9.

Response patterns for the four alcohols are shown in Fig. 5.9;[24] we can see that four different patterns are obtained. However, discrimination of the alcohols was impossible using only metal oxide gas sensors. Measurement of the alcohol concentrations was attempted using the neural network algorithm. As a result, the identification of the different alcohols was almost satisfactory for the concentration range 0–1000 ppm. If the three metal oxide gas sensors were removed from the network, it was not possible to identify all of the alcohols.

An array of gas-sensitive field-effect devices and metal oxide gas sensors was applied to the identification of cheeses and classification of ground meat. Samples were stored in 400 ml glass beakers covered with parafilm. After the storage time, the vapor in each beaker was passed to the sensor array for 60 s. Using a neural network, identification of cheeses was successfully achieved. Ground pork and ground beef with different storage times was differentiated using the CO_2 monitor together with the two types of sensor used in the above array. The CO_2 monitor was effective to identify beef with different storage days. These sensors were used satisfactorily over a period of 10–30 days. Development of calibration procedures will increase the feasibility of these sensors.

The response patterns (i.e., "olfactory images") formed by molecules interacting with gas-sensitive field-effect structures with large area were demonstrated using a scanning light pulse technique (Fig. 5.10). This

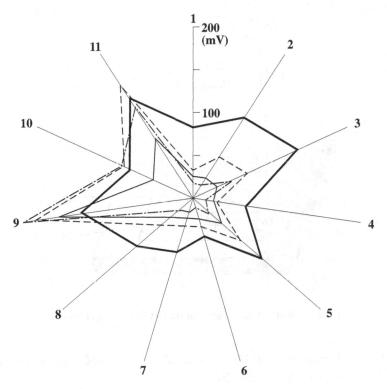

Figure 5.9. Response patterns for methanol (━━), ethanol (- - - - -), 1-propanol (———) and 2-propanol (— - —).[24] The sensors were subjected to 100 ppm of the different alcohols in air for 15 min. Ch. 1: FET 100 nm Ir at the operating temperature 150 °C; ch. 2: FET 100 nm Pd at 150 °C; ch. 3: FET 100 nm Ir/10 nm Pt at 170 °C; ch. 4: FET 100 nm Ir at 180 °C; ch. 5: FET 50 nm Ir at 190 °C; ch. 6: FET 60 nm Pd at 150 °C; ch. 7: FET 100 nm Pt/10 nm Fe at 160 °C; ch. 8: FET 350 nm Pd at 195 °C; ch. 9: Figaro TGS 813; ch. 10: Figaro TGS 800; ch. 11: Figaro TGS 881.

method generates considerable information, comparable to that obtained from an array comprising many sensors. If the voltage is kept constant, the photocurrent can be recorded as a function of the position of the light spot on the sensor surface that was placed under the temperature gradient. While image processing and presentation of data is now in progress, the method is effective for distinguishing odorants easily because the human eye directly observes the subtle differences in the obtained patterns.

The gas sensor array has also been applied to monitoring biopharmaceutical processes.[25,26] The characteristic transitions during the main growth or production phases were visualized in two recombinant bioprocesses, the

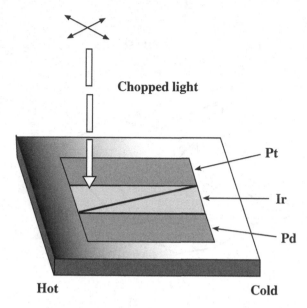

Figure 5.10. Illustration of the scanning light pulse technique.[22,24]

production of human growth hormone in *Escherichia coli* and human factor VIII in Chinese ovary hamster cells.

5.6 Odor sensor using a surface acoustic wave device

A surface acoustic wave (SAW) device in a radiofrequency circuit can function as a sensor of the mass adsorption of the device, where the frequency change of the circuit is proportional to the mass of adsorption, as with a quartz oscillator. SAW is a kind of mechanical, transverse wave generated in the elastic body when it is patted or a forced mechanical oscillation is applied to it.

A SAW device with a Langmuir–Blodgett (LB) film for odor sensing has been developed.[27] The main principle and structure are almost the same as that in the odor sensor utilizing QCM described in Section 5.2. Arachidic acid, methyl arachidate, icosanol, tricosanoic acid, 22-tricosenoic acid and 22-tricosynoic acid were used as amphiphilic materials, and a four-layered LB film was used.

The mass of adsorption of an odor molecule was plotted against the hydrophobicity of the amphiphilic material, i.e., the partition coefficient in the 1-octanol/water system. As a result, it was found that the regression line for the three materials (arachidic acid, methyl arachidate, icosanol) with 20 carbons in the hydrocarbon chain was different from that of the materials (tricosanoic acid, 22-tricosenoic acid, 22-tricosynoic acid) with 23 carbons; adsorption of odor molecules was greater to the materials made with the 20 carbon chains than for those with 23 carbons. It may be related to the rigidness of the membrane structure. Odorants are supposed to adsorb to hydrocarbon chains which are in a more fluid state.

This system was used to attempt to distinguish seven fragrances. The group of fragrances from plants (Floral 1, Floral 2, Green and Citrus) was clearly discriminated from the group from animals (Oriental, Aldehydic and Musky-up). The PCA showed that the seven fragrances were clearly separated and the discrimination of the fragrances was successfully achieved. The gas-sensitive device with the 22-tricosynoic acid LB film made the greatest contribution to the recognition of the fragrances.

The result of this odor recognition system was similar to the human discrimination of odor of gases, i.e., average individuals could discriminate the group of fragrances from plants from the group from aminals at first. Floral 1 and Floral 2, which have similarly perceived odor, were located closely in the two-dimensional plane of PCA applied on the sensor output. This biomimetic property may originate from the use of the LB membrane, which partly has a structure and materials similar to those found in biological systems.

5.7 Detection of odorants using monolayer membranes

The sensors described so far respond to both water vapor or flammable gases, such as methane, ethane and hydrogen gas, that show low solubility in water and have no smell. Therefore, no artificial system to execute the sense of olfaction has been completed. A possible reason for this could be that these sensors do not mimic the biological olfactory system, where chemical substances which show odor must be soluble in the water phase in the nasal cavity. The substances then reach the olfactory cells by diffusion. This means that odorants must have some hydrophilic properties as well as hydrophobic properties. In the recognition of odor, there is a possibility that this characteristic plays an important role. Based on this idea, detection of odorants in water phase was tried using an electrochemical method, cyclic voltammetry.[28]

Thiol-containing compounds form monolayer membranes on a gold surface via chemisorption from organic solvents because of a strong affinity between thiol and the metal.[29,30] Here we prepared different kinds of

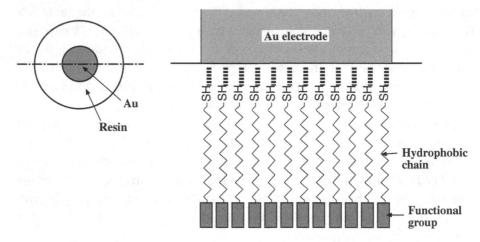

Figure 5.11. Bottom surface (left) and cross-section (right) of a gold electrode modified with a monolayer of a thiol-containing lipid.[28]

thiol-containing lipid and fabricated monolayer membranes on gold disk electrodes.

The monolayer-membrane electrode of a thiol-containing lipid (Fig. 5.11) was formed according to the following procedure. A gold disk electrode of diameter 1.6 mm was polished with alumina powder (particle size 0.3 μm), and then immersed into an ethanoic solution of a thiol-containing compound at room temperature. Five different thiol-containing lipid materials were used: $HS(CH_2)_{10}COOH$, $[HS(CH_2)_{11}O]_2POOH$, $HS(CH_2)_{20}N^+(CH_3)_3Br^-$, $HS(CH_2)_2COOH$ and $HS(CH_2)_{17}CH_3$.

The cell was sealed hermetically, and nitrogen gas was used for deaerating the experimental solution. The monolayer electrode (working electrode), a reference electrode composed of an Ag/AgCl wire with saturated KCl, and a counter electrode of platinum wire were all connected to the potentiostat. The voltage difference between the monolayer electrode and reference electrode was swept by the potentiostat and a function generator. Cyclic voltammograms were obtained by recording the applied voltage and the electric current between the counter electrode and the monolayer electrode. Some peaks of electric current appeared and the values were determined by the concentration and diffusion coefficients of redox substances. Permeability changes could then be estimated from waveform changes of the cyclic voltammograms.

Figure 5.12 shows the response patterns for three odorants, β-ionone, β-citral and chloroform, where the peak current normalized by the initial peak current in the cyclic voltammogram was taken as the response. These

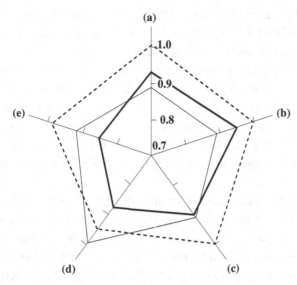

Figure 5.12. Response patterns for three odorants.[28] The response is the peak current normalized by the initial peak current. (a) $HS(CH_2)_2COOH$, (b) $HS(CH_2)_{10}COOH$, (c) $[HS(CH_2)_{11}O]_2POOH$, (d) $HS(CH_2)_{20}N^+(CH_3)_3Br^-$, (e) $HS(CH_2)_{17}CH_3$. $2\,\mu M$ β-ionone (———), $20\,\mu M$ β-citral (———), and $1\,mM$ chloroform (- - - - -).

odorants are easily discriminated. The threshold values of the human olfactory system are about 10^{-4} mol/l for chloroform, 10^{-7} mol/l for β-citral and 10^{-10} mol/l for β-ionone in the vapor phase. The lowest response thresholds among the five monolayer electrodes were about 10^{-4} M for chloroform, 10^{-7} M for β-citral and 10^{-8} M for β-ionone. The order of the threshold values is similar to that of the human olfactory sense. This result may imply that the responses of the monolayer membranes resemble those of the biological membranes in odor reception.

All five membranes responded to β-ionone at very low concentrations. The membrane made of $HS(CH_2)_{17}CH_3$ began to respond to β-ionone at a lower concentration than those made from the other lipids. It is considered that β-ionone acted on the hydrophobic chain and raised the packing density of the membrane because β-ionone has a benzene nucleus and a large molecular structure.

Chloroform has a low molecular weight, a small molecular structure and a weak hydrophobicity compared with other odor substances used in this experiment. Therefore, it seems that chloroform hardly acts on the monolayer membrane and would not be helpful in changing the redox-blocking ability of the membrane. The membranes with the negatively charged groups of $-COO^-$ and $-POO^-$ hardly responded to chloroform, and the

$HS(CH_2)_{17}CH_3$ membrane responded to chloroform only slightly. The $HS(CH_2)_{20}N^+(CH_3)_3Br^-$ membrane responded to chloroform profoundly. It is, therefore, considered that the greater the positive charge of the functional group of lipids becomes, the greater is the response to chloroform.

This sensing system may have a profound ability for odor sensing because of the following three factors: (a) the simplicity of forming a lipid monolayer membrane, (b) the stability of the coordination between gold and thiol, and (c) the similarity of the odor-receiving mechanism utilizing biomaterials such as lipids and water phase to the actual human olfactory system.

Properties of the lipid materials such as the length of the hydrophobic chain and the kind of functional group used for forming the monolayer membrane are also important for the responses to odorants. The responses of the membrane to odor are determined by the adsorbing manner of the odor substance, which has various properties such as the balance of hydrophilicity and hydrophobicity strength, the molecular weight, the molecular structure and the functional group. Considering these factors, we can expect to observe many different characteristic responses to odorants with the use of various lipid materials.

The same method was applied to detection of taste substances in liquids.[31] Electrolytic taste substances promoted and nonelectrolytic substances suppressed the permeability of the membrane. The order of the thresholds detected was quinine < HCl < caffeine < MSG < NaCl < sucrose, in accordance with that found in humans.

A recent study shows quantitative detection of aging processes of milk and orange juice stored at room temperature using pulse voltammetry, where successive voltage pulses of gradually changing amplitudes are applied to the working electrodes.[32] Gold and platinum were used as the working electrodes, and PCA was applied on the obtained data. This method may be powerful in detection of small changes in quality of foods because of its high sensitivity and resolution.

Göpel *et al.* proposed a new affinity biosensor using a quartz oscillator coated with self-assembled thiols as a selective monolayer.[33,34] A synthetic peptide (epitope as antigen) was modified with ω-hydroxyundecanethiol (HUT) by a succinyl linker (HUT-peptide). The monolayer of HUT and HUT-peptide was self-assembled on gold. The immunological reaction, where the peptide is recognized by specific antibodies, was observed by measuring the resonance frequency change of QCM. This method using QCM coated with thiols has considerable potential as an electronic nose by choosing suitable coating materials.

REFERENCES

1. Persaud, K. C. and Dodd, G. H. (1982). *Nature*, 299, 352.
2. Kashiwayanagi, M. and Kurihara, K. (1984). *Brain Res.*, 293, 251.
3. Kurihara, K., Yoshii, K. and Kashiwayanagi, M. (1986). *Comp. Biochem. Physiol.*, 85A, 1.
4. Moriizumi, T., Nakamoto, T. and Sakuraba, Y. (1994). In *Olfaction and Taste XI*, eds. Kurihara, K., Suzuki, N. & Ogawa, H. Springer-Verlag, Tokyo, p. 694.
5. Ema, K., Yokoyama, M., Nakamoto, T. and Moriizumi, T. (1989). *Sens. Actuators*, 18, 291.
6. Nakamoto, T., Sasaki, S., Fukuda, A. and Moriizumi, T. (1992). *Sens. Mater.*, 4, 111.
7. Okahata, Y., Shimizu, O. and Ebato, H. (1990). *Bull. Chem. Soc. Jpn*, 63, 3082.
8. Nakamoto, T., Fukuda, A., Moriizumi, T. and Asakura, Y. (1991). *Sens. Actuators*, B3, 221.
9. Nakamoto, T., Fukuda, A. and Moriizumi, T. (1993). *Sens. Actuators*, B10, 85.
10. Nanto, H., Tsubakino, S., Ikeda, M. and Endo, F. (1995). *Sens. Actuators*, B24-25, 794.
11. Hierlemann, A., Weimar, U., Kraus, G., Gauglitz, G. and Göpel, W. (1995). *Sens. Mater.*, 7, 179.
12. Matsuno, G., Yamazaki, D., Ogita, E., Mikuriya, K. and Ueda, T. (1995). *IEEE Trans. Instru. Measurement*, 44, 739.
13. Di Natale, C., Brunink, J. A. J., Bungaro, F., Davide, F., D'Amico, A., Paolesse, R., Boschi, T., Faccio, M. and Ferri, G. (1996). *Meas. Sci. Technol.*, 7, 1103.
14. Bodenhöfer, K., Hierlemann, A., Seemann, J., Gauglitz, G., Koppenhoefer, B. and Göpel, W. (1997). *Nature*, 387, 577.
15. Bartlett, P. N. and Gardner, J. W. (1992). In *Sensors and Sensory Systems for an Electronic Nose*, eds. Gardner, J. W. & Bartlett, P. N. Kluwer Academic, the Netherlands, p. 31.
16. Gardner, J. W. and Bartlett, P. N. (1994). In *Olfaction and Taste XI*, eds. Kurihara, K., Suzuki, N. & Ogawa, H. Springer-Verlag, Tokyo, p. 690.
17. Gardner, J. W., Shurmer, H. V. and Tan, T. T. (1992). *Sens. Actuators*, B6, 71.
18. Nanto, H., Tsubakino, S., Kawai, T., Ikeda, M., Kitagawa, S. and Habara, M. (1994). *J. Mater. Sci.*, 29, 6529.
19. Di Natale, C., Davide, F. A. M., D'Amico, A., Nelli, P., Groppeli, S. and Sberveglieri, G. (1996). *Sens. Actuators*, B33, 83.
20. Persaud, K. C. and Pelosi, P. (1992). In *Sensors and Sensory Systems for an Electronic Nose*, eds. Gardner, J. W. & Bartlett, P. N. Kluwer Academic, the Netherlands, p. 237.
21. Dall'Olio, A., Dascola, G., Varacca, V. and Bocchi, V. (1968). *C. R. Acad. Sci., Paris Ser. C*, 267, 433.
22. Lundström, I., Erlandsson, R., Frykman, U., Hedborg, E., Spetz, A., Sundgren, H., Welin, S. and Winquist, F. (1991). *Nature*, 352, 47.
23. Lundström, I., Ederth, T., Kariis, H., Sundgren, H., Spetz, A. and Winquist, F. (1995). *Sens. Actuators*, B23, 127.
24. Winquist, F., Sundgren, H., Hedborg, E., Spetz, A., Holmberg, M. and Lundström, I. (1992). *Tech. Digest 11th Sens. Symp.*, p. 257.
25. Mandenius, C.-F., Hagman, A., Dunas, F., Sundren, H. and Lundström, I. (1998). *Biosens. Bioelectron.*, 13, 193.
26. Lidén, H., Mandenius, C.-F., Gorton, L., Meinander, N. Q., Lundström, I. and Winquist, F. (1998). *Anal. Chim. Acta*, 361, 223.
27. Ohnishi, M., Ishibashi, T., Kijima, Y., Ishimoto, C. and Seto, J. (1992). *Sens. Mater.*, 4, 53.
28. Miyazaki, Y., Hayashi, K., Toko, K., Yamafuji, K. and Nakashima, N. (1992). *Jpn J. Appl. Phys.*, 31, 1555.
29. Porter, M. D., Bright, T. B., Allara, D. L. and Chidsey, C. E. D. (1987). *J. Am. Chem. Soc.*, 109, 3559.

30. Nakashima, N., Takada, Y., Kunitake, M. and Manabe, O. (1990). *J. Chem. Soc.,*
    Chem. Commun., 1990, 845.
31. Iiyama, S., Miyazaki, Y., Hayashi, K., Toko, K., Yamafuji, K., Ikezaki, H. and
    Sato, K. (1992). *Sens. Mater.*, 4, 21.
32. Winquist, F., Wide, P. and Lundström, I. (1997). *Anal. Chim. Acta*, 357, 21.
33. Göpel, W. and Heiduschka, P. (1995). *Biosens. Bioelectron.*, 10, 853.
34. Rickert, J., Weiss, T., Kraas, W., Jung, G. and Göpel, W. (1996). *Biosens. Bioelectron.*,
    11, 591.

6

Taste sensors

———

A multichannel taste sensor, i.e., electronic tongue, with global selectivity is composed of several types of lipid/polymer membrane for transforming information of taste substances into electric signals, which are input to a computer. The sensor output shows different patterns for chemical substances that have different taste qualities such as saltiness and bitterness, while it shows similar patterns for chemical substances with similar tastes. The sensor responds to the taste itself, as indicated by the fact that taste interactions such as the suppression effect, which occurs between sweet and bitter substances, can be reproduced well. Amino acids can be classified into several groups according to their own tastes based on sensor outputs. The mixed taste elicited by amino acids such as L-valine and L-methionine, which implies that the taste comprises two or three taste qualities such as sweetness and bitterness, is expressed quantitatively by a combination of the basic taste qualities elicited by L-alanine and L-tryptophan. The taste of foodstuffs such as beer, coffee, mineral water, milk, sake, rice, soybean paste and vegetables can be discussed quantitatively using the taste sensor, which provides the objective scale for the human sensory expression. Miniaturization of the taste sensor using FET gives a sensor with the same characteristics as the above taste sensor by measuring the gate-source voltage. The taste sensor can also be applied to measurements of water pollution. It will open doors to a new era of food and environmental sciences.

6.1 Measurement of taste

One of the goals of sensor technology is to reproduce the five types of sense seen in humans or to surpass their abilities. To do this, sensors must have three characteristics: high sensitivity, stability and high selectivity. The first two items seem to be fairly reasonable. However, the last is not always achieved.

The reception of light in the sense of sight, sound in hearing and pressure or temperature in touch all involve physical quantities that need to be detected with high sensitivity and selectivity. Sensors to detect physical quantities are called physical sensors. These sensors are made so as to detect only one physical quantity with high selectivity.

For example, let us consider the situation where we measure the body temperature when we are sick. In this case, we set a thermometer to our mouth or under the arm, where it is dark. If the thermometer responds to the reduction in light, it is quite troublesome, i.e., the thermometer should measure only the temperature. The same situation also applies to the microphone. It should respond only to the voice and should not respond to light and the other quantities. These examples confirm that the sensors reproducing the senses of sight, hearing and touch should respond to only the one physical quantity to be measured.

However, the situation is different in the senses of taste and smell. For example, over 1000 chemical substances are included in tea or coffee. Humans do not distinguish each chemical substance. Instead, humans decompose the taste of foodstuffs into five types of basic taste quality: sour, bitter, salty, sweet and umami. This implies that we need to develop a novel type of sensor for taste and smell.

Taste is constructed from five basic taste qualities as above: sourness produced by H^+ of HCl, acetic acid, citric acid, etc.; saltiness produced mainly by NaCl; bitterness produced by quinine, caffeine and $MgCl_2$; sweetness by sucrose, glucose, etc.; and umami taste, produced by MSG, contained mainly in seaweeds, IMP, occurring in meat and fish and GMP, found in mushrooms (see Chapter 2).

Conventional chemical sensors for reproducing the chemical properties of biological systems have been developed to be selective for chemicals, as mentioned in Chapter 4. Those biosensors are not competent to sense taste, because it is hardly possible to assemble selective sensors for all the chemical substances involved in the production of taste. Furthermore, the taste interaction cannot be reproduced using these kinds of selective biosensor. Whereas the conventional biosensors have the same concept, i.e., selectivity, as physical sensors, a new type of sensor is required to measure taste. The intention is not to measure the amount of each chemical substance but to measure the taste itself, and to express it quantitatively.

In biological taste reception, taste substances are received by the biological membrane, which is composed of lipids and proteins, of gustatory cells in taste buds on the tongue (see Chapter 2). Then the information on taste substances is transduced into an electric signal that is transmitted along the nerve fiber to the brain, where the taste is perceived.

As shown in Chapter 3, lipid membranes can respond to taste substances and can detect interactions between taste substances in the same way as occurs in biological taste. These results indicate that a taste sensor can be realized by the use of lipid membranes as transducers.

A multichannel electronic taste-sensing system, i.e., electronic tongue, utilizes a transducer composed of lipid/polymer membranes. This sensor can detect tastes in a manner similar to human gustatory sensation. The output of the sensor is not the amount of specific molecules to exhibit taste but the taste quality and intensity, because different output electric patterns were obtained for different taste groups such as sourness and saltiness. However, similar patterns were obtained for molecules in the same taste groups, such as MSG, IMP and GMP, which have an umami taste, and NaCl, KCl and KBr for saltiness.

Discrimination of each chemical substance is not important here, but recognition of the taste itself and its quantitative expression must be made. The taste sensor using lipid/polymer membranes is based on a concept of global selectivity, which implies the ability to classify enormous numbers of chemical substances into several groups, as really found in the taste reception in biological systems.

Quantification of taste is possible using the taste sensor, and hence we can discuss the taste objectively.

6.2 Multichannel taste sensor

Transducers of the multichannel taste sensor were composed of lipid membranes immobilized with a polymer.[1-3] Eight lipid analogs were used for preparing the membranes, as summarized in Table 6.1 including the mixed, hybrid membranes composed of dioctyl phosphate (DOP) and trioctyl methyl ammonium chloride (TOMA). The major parts of the functional groups in the biological membranes are lined up with these lipids. The lipid materials used varied with the substances to be measured. In the subsequent discussion we will not indicate which particular lipid was used in each application.

Each lipid was mixed in a test tube with polyvinyl chloride (PVC) and a plasticizer (dioctyl phenylphosphonate, DOPP) dissolved in tetrahydrofuran. Successively, the mixture was dried on a glass plate, which was placed on a hot plate controlled at approximately 30 °C. The lipid/polymer membrane

Table 6.1. *Lipid materials used in the multichannel electrode*

Channel	Lipid (abbreviation)
1	*n*-Decyl alcohol (DA)

$$HO-(CH_2)_9CH_3$$

| 2 | Oleic acid (OA) |

$$HO-\overset{\overset{\displaystyle O}{\|}}{C}-CH_2(CH_2)_7CH=CH(CH_2)_7CH_3$$

| 3 | Dioctyl phosphate (bis(2-ethylhexyl)hydrogen phosphate, DOP) |

$$HO-\overset{\overset{\displaystyle O}{\|}}{P}\overset{\diagup OCH_2\overset{\overset{\displaystyle C_2H_5}{|}}{CH}(CH_2)_3CH_3}{\diagdown OCH_2\underset{\underset{\displaystyle C_2H_5}{|}}{CH}(CH_2)_3CH_3}$$

4	DOP : TOMA 9 : 1
5	DOP : TOMA 5 : 5
6	DOP : TOMA 3 : 7
7	Trioctyl methyl ammonium chloride (TOMA)

$$\underset{CH_3}{\overset{(CH_2)_7CH_3}{{}^+N{<}(CH_2)_7CH_3}}\;{(CH_2)_7CH_3}$$

| 8 | Oleyl amine (OAm) |

$$\overset{H}{\underset{H}{>}}N-CH_2(CH_2)_7CH=CH(CH_2)_7CH_3$$

thus prepared was a transparent, colorless and soft film of approximately 200 μm thickness. The surface structure of a lipid/polymer membrane is illustrated in Fig. 6.1. The hydrophilic groups of the lipids contact the aqueous phase, while the hydrophobic hydrocarbon chains are buried in the membrane matrix.

The membrane made of PVC and DOPP shows selectivity to cations;[4] however, the mechanism by which this occurs is not clear because this membrane should have no negative electric charge to cause cation selectivity. Recently, identification of the charged impurity phenylphosphonic acid monooctyl ester in commercially available DOPP was made using FRIT fast atom bombardment (FRIT-FAB).[5] The membrane made of equimolar DOP and TOMA is slightly negatively charged because of the negatively charged impurity.

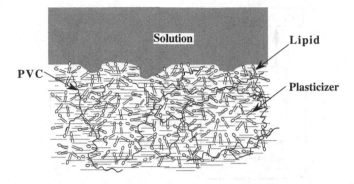

Figure 6.1. Surface structure of a lipid/polymer membrane.

Lipid/polymer membranes were fitted on a multichannel electrode. The detecting electrode of each channel (ch.) was made up of silver wires surface plated with Ag/AgCl, which were embedded in a basal acrylic board of 2 mm thickness, as shown in Fig. 6.2. Another acrylic board of 1 cm thickness with eight cone-shaped holes was affixed to this board. The holes were filled with 100 mM (or 3 M) KCl solution, and the eight membranes were fitted on the board to cover the holes.

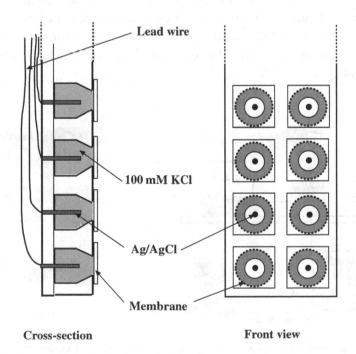

Figure 6.2. Multichannel electrode with eight lipid/polymer membranes.

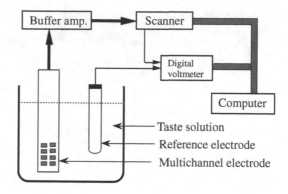

Figure 6.3. Measurement system for the laboratory use.

The multichannel electrode was connected to an 8-channel scanner through high-input impedance amplifiers (Fig. 6.3). The selected electric signal from the sensor was converted to a digital code by a digital voltmeter and was fed to a computer. Then, the voltage difference between the multichannel electrode and an Ag/AgCl reference electrode was measured.

The commercial taste-sensing systems (SA401, SA402, Anritsu Corp., Figs. 6.4(*a*) and (*b*), respectively) have essentially the same mechanism as that described above. The detecting sensor part is made of seven (or eight) electrodes of lipid/polymer membranes and is controlled mechanically by a robot arm. Measurements of the tastes of five chemical substances that produce basic taste and of Japanese sake, coffee and milk (see below) were made using SA401.

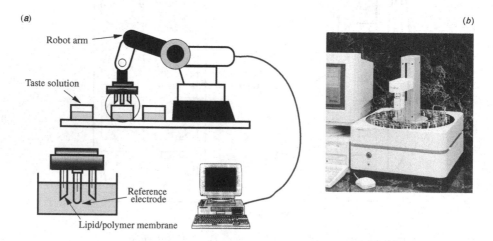

Figure 6.4. Commercially available taste-sensing systems (Anritsu Corp.).
(*a*) SA401, (*b*) SA402.

The measurements were usually made three times by the rotation procedure, which involved three rounds of measurement of all the samples. The measurement time was 20–60 s, dependent on the samples.

6.3 Response characteristics

6.3.1 Five basic taste qualities

Five chemical substances producing the five typical tastes—HCl (sour), NaCl (salty), quinine-HCl (bitter; hereafter called quinine for simplicity), sucrose (sweet) and MSG (umami taste)—were studied.[1,6] The results[6] are shown in Fig. 6.5.

The starting origin (0 mV) is chosen to be the response electric potential at 1 mM KCl with no added taste-producing chemical substance. Standard deviations in the same eight membranes were small: for example, in the membrane comprising DOP : TOMA 5 : 5, the deviations were 0.18, 0.43, 0.56, 0.77, 0.36, 0.76, 0.89, 0.96 mV for 0.001, 0.003, 0.01, 0.03, 0.1, 0.3, 1.0, 10.0 mM quinine, respectively. The response electric potential is different

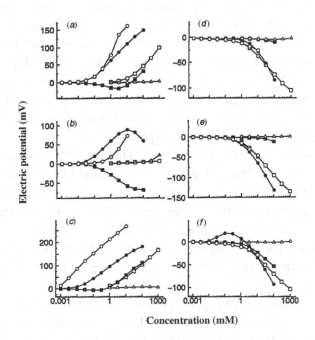

Figure 6.5. Responses of sensors comprising different lipid membranes (*a–f*, see Table 6.1) to five basic taste substances. (*a*) DOP; (*b*) DOP : TOMA 5 : 5; (*c*) OA; (*d*) DOP : TOMA 3 : 7; (*e*) TOMA; (*f*) OAm. ●, HCl; □, NaCl; ○, quinine; △, sucrose; ■, MSG.

for different chemical substances in each membrane and, furthermore, is different in other membranes.

For NaCl, ch. 3 (DOP) responds upward from about 5 mM (Fig. 6.5(*a*)), and ch. 5 (DOP:TOMA 5:5) shows no response (Fig. 6.5(*b*)), whereas ch. 7 (TOMA) responds downward from about 1 mM (Fig. 6.5(*e*)). The recognition threshold of humans is 30 mM;[7] hence the sensitivity of the sensor is superior to that of humans. As discussed in Section 6.6, the upward and downward responses are caused by Na^+ and Cl^-, respectively, because chs. 3 and 7 are negatively charged and positively charged membranes, respectively. It is noticeable that the response electric potentials in chs. 1–3 do not become saturated with increasing NaCl concentration (see chs. 3 and 2 in Fig. 6.5(*a*) and (*c*), respectively). This reason will be clarified in Section 6.6.1.

For HCl, too, a similar situation holds except for the threshold; for example, about 0.05 mM HCl induces the changes in response electric potential in ch. 3 in Fig. 6.5(*a*). This value is much lower than the human recognition threshold, 0.9 mM. In responses to quinine, the high sensitivity as represented by the threshold of 3 μM can be attained in chs. 1 (*n*-decyl alcohol (DA) and 2 (oleic acid (OA)), while it is 30 μM in humans.

For MSG, chs. 2 and 3 show clear biphasic responses, with an increase following a decrease in electric potential. By comparison, the response electric potential of ch. 5 (DOP:TOMA 5:5) decreases continuously (Fig. 6.5(*b*)). These results are brought about by negatively charged glutamate ions of MSG, which are bound to the hydrophilic part of lipid to induce the downward response, and Na^+, which induces the upward response. The continuous decrease occurs because this membrane does not respond to Na^+. MSG causes salty and umami taste sensations simultaneously in a human.[8,9] The responses in chs. 2 and 3 reproduce this phenomenon.

Channels 2 (OA), 3 (DOP) and 5 (DOP:TOMA 5:5) respond upward to quinine, while chs. 6 (DOP:TOMA 3:7), 7 (TOMA) and 8 (oleyl amine (OAm)) respond downward. The upward response is brought about by adsorption of quinine ions to the membrane, while the downward response results from the electric screening effect by Cl^-.

The potential does not change significantly through chs. 1 to 8 by the application of sucrose. The potential slightly increases above 30 mM sucrose in all channels.

As shown in Section 6.6, changes in electric charge density of the membrane surface and/or ion distribution near the surface are caused by taste-producing substances such as HCl, NaCl, quinine and MSG, which are electrolytes. However, a large change in the ionic distribution may not occur with sucrose, since sucrose is a nonelectrolyte. Sucrose may be received with a conformational change in the lipid membrane or a slight neutralization of the head groups of lipid molecules by the weak dipole of a sucrose molecule, which leads to the change in the surface electric charge density to cause the change in membrane potential.

The response electric potential is different for chemical substances showing different taste qualities in each membrane and, furthermore, is different in other membranes. This implies that taste quality is distinguishable using the response pattern constructed from response electric potentials of several membranes. However, the taste sensor has similar response patterns to the same group of taste; for example the sour substances HCl, citric acid and acetic acid show similar response patterns. Salty substances, NaCl, KCl and KBr, show similar patterns, too, as shown in Fig. 6.6. The response patterns bulge to the upper-right direction, whereas the responses of the DOP : TOMA 5 : 5 membrane are nearly zero. The umami taste substances MSG, IMP and GMP all give similar patterns. In almost all the membranes, the responses are negative. The patterns are clearly different from those for salty substances. MSG is amino acid, while IMP and GMP are nucleotides. In spite of large difference of chemical structure between these molecules, the taste sensor has similar patterns. Therefore, we can conclude that this taste sensor can respond to the taste itself.

Figure 6.7 shows a three-dimensional taste map that was obtained by application of PCA to the response patterns. It contains HCl, citric acid and acetic acid for sourness, NaCl, KCl and KBr for saltiness, quinine, L-trypto-phan and L-phenylalanine for bitterness, sucrose, glucose and fructose for

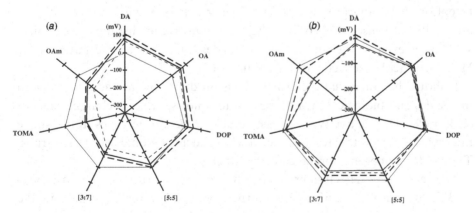

Figure 6.6. Responses of different membranes to taste substances. (*a*) Salty substances (300 mM): NaCl (▬ ▬ ▬); KBr (- - - - -); KCl (▬▬▬). (*b*) Umami substances (1 mM): MSG (▬ ▬ ▬); IMP (- - - - -); GMP (▬▬▬).

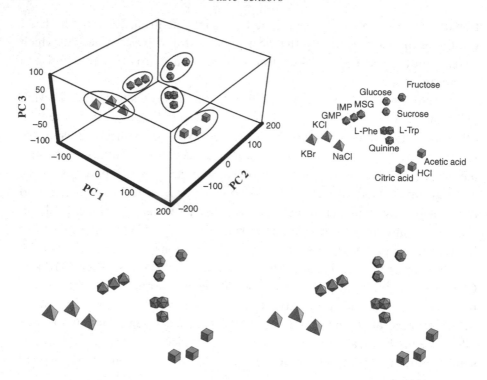

Figure 6.7. Three-dimensional taste map of five types of taste quality obtained by application of the PCA to the response electric potential patterns. Simultaneously following the lower two figures makes a stereograph. The units for PC1, PC2 and PC3 are mV.

sweetness, and MSG, IMP and GMP for umami taste. We can see clear grouping of five taste qualities in the three-dimensional space.

This implies that the taste sensor has global selectivity, as in the taste reception in biological systems. This sensor can be regarded essentially as an intelligent sensor. Of course, the taste sensor can distinguish chemical substances even in the same group of taste; for example, the patterns for HCl, citric acid and acetic acid are different.

It should be noted that the taste sense is produced by interactions between molecules and biological membranes: taste is not specific or characteristic of each molecule. The five types of taste quality reflect differences in the above interactions. From this reason, the situation such as global selectivity arises. The lipid/polymer membrane can reproduce it.

The sensor had detection errors (in the unit of logarithmic concentration) 0.73% for saltiness, 0.65% for sourness and 3.4% for bitterness in the aqueous solution containing simultaneously different chemical substances producing different taste qualities.[10] Humans usually cannot distinguish two tastes with a concentration difference below 20%.[7] Here, 20% means an error of 7.9% (log 1.2). Therefore, the detection ability of the sensor is superior to that of humans.

6.3.2 Pungency and astringency

Knowledge of taste sensation is particularly lacking in qualities called pungency and astringency, because these are not basic tastes.

Capsaicin, the primary pungent agent of hot red pepper, has been empirically supposed to induce a warming action. Pharmacological studies showed that capsaicin produced its warming effect by promoting the secretion of catecholamine from the adrenal medulla.[11] The warming effect of capsaicin occurs gradually, while it has another rapid "pungent" effect, the mechanism of which is not known. The pungent substances such as capsaicin, piperin and allyl isothiocyanate offer a violent stimulation to the oral cavity.

Astringent substances such as tannic acid are bound to proteins and precipitate them. They are most effective at pH values near their isoelectric points. Tightly coiled globular proteins have much lower affinities for tannic acid than conformationally loose proteins.[12,13] This phenomenon has long been recognized as being responsible for the astringency of unripe fruit and various beverages. For this reason, Bate-Smith suggested that astringency is not a taste sense but a touch sense.[14] In contrast, some studies showed that oral astringency is closely linked to bitterness.[15,16] Moreover, clear electrophysiological responses to tannic acid were obtained from rat chorda tympani and lingual nerves.[17,18] These results imply that astringency is also a chemical taste.

The response of the taste sensor to capsaicin was studied.[19] Capsaicin, piperine and allyl isothiocyanate had no effect on the electric potentials of lipid/polymer membranes of the taste sensor. These results agree with pharmacological observations. These substances will not produce a chemical taste as widely recognized in the field of taste sensation. Some reports[20,21] suggest the presence of capsaicin-sensitive neurons, which are associated with taste buds, separated from primary taste neurons.

The effects of tannic acid, catechin, gallic acid and chlorogenic acid on the taste sensor were investigated.[19] The pH of these solutions at 1 mM were 3.93, 5.68, 3.27 and 3.61, respectively. The response potential of lipid/polymer membranes of the taste sensor to tannic acid is shown in Fig. 6.8. The astringent substances caused marked changes in the membrane potentials of the sensor. The relative standard deviations with five measurements on each taste substance were 1 to 2%.

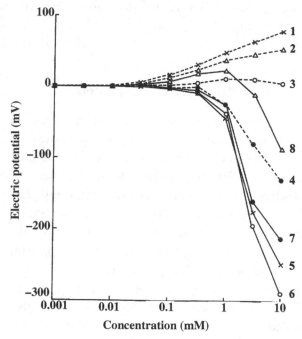

Figure 6.8. Response potentials of the taste sensor to tannic acid in chs. 1–8 in Table 6.1.[19]

The responses to tannic acid and gallic acid did not resemble each other; the membranes with positively charged lipids such as DOP:TOMA 3:7 and TOMA responded well to tannic acid, while the membranes with negatively charged lipids such as DOP, DA and OA responded well to gallic acid. The changes of the membrane potentials induced by tannic acid were larger than those by gallic acid. However, the threshold concentrations to alter the membrane potentials are about 0.3 mM for tannic acid and 0.01 mM for gallic acid. Catechin resembled tannic acid, while chlorogenic acid resembled gallic acid in the response pattern.

The taste sensor with lipid membranes responded to astringent substances. Hence, we searched for qualities of taste similar to astringency. Lyman and Green suggested that astringency is closely linked to bitterness.[16] In fact, a response pattern similar to that of tannic acid was obtained with a bitter substance, sodium picrate. Using the PCA of the patterns obtained with the taste sensor, it was found that astringent substances are located near the bitter substance (sodium picrate). The qualities of tannic acid and catechin are close to sodium picrate, whereas those of gallic acid and chlorogenic acid are located in the midst of qualities of sodium picrate and HCl.

The astringent substances, which are well known to generate a tactile sensation,[14,22] altered the response electric potentials of the taste sensor markedly.

The naturally occurring phenolic compounds that dissociate into ions in solution are correlated with astringency. Figure 6.8 shows that the negatively charged portion of tannic acid would have affected the positively charged lipids such as DOP:TOMA 3:7 and TOMA to lower the membrane potential. Dissociated protons would have affected the negatively charged lipids such as DOP, DA and OA to increase the membrane potential.

These results indicate that astringency is related to both the senses of bitter taste and touch.

6.4 Taste of amino acids

Taste of amino acids was studied using the taste sensor.[3,23-25] Each of amino acids elicits complicated mixed taste itself, as shown in Table 2.2 (p. 34); for example, L-valine produces sweet and bitter tastes at the same time. There are detailed data on taste intensity and taste quality of various amino acids from sensory panel tests.[26-28] Amino acids are generally classified into several groups that correspond to each characteristic taste. The study described below is the first trial to study the taste of amino acids using artificial sensing devices. The lipids used are detailed in Table 6.1.

6.4.1 Classification of taste of amino acids

Figure 6.9 shows the response patterns to typical amino acids, each of which elicits different taste quality in humans. Each channel responded in different ways depending on the taste of the amino acid. L-Tryptophan, which elicits almost pure bitter taste in humans, increased the potentials of chs. 1, 2 and 3 greatly. Typical examples of standard deviations were 3.16, 7.32, 1.12, 2.51, 2.37, 2.86, 1.31 and 2.33 mV for chs. 1–8, respectively. This tendency was also observed for other amino acids that mainly exhibit bitter taste, such as L-phenylalanine and L-isoleucine. L-Valine and L-methionine, which taste mainly bitter and slightly sweet, decreased the potential of ch. 5; the responses of chs. 1 and 2 were small. For L-alanine, glycine and L-threonine, which taste mainly sweet, the potentials of chs. 1 and 2 decreased.

L-Glutamic acid and L-histidine monohydrochloride, which taste mainly sour, increased each of the potentials of chs. 1–5 to almost the same degree. Monosodium L-aspartate, eliciting mainly umami taste, showed a response pattern that differed from those of the other amino acids.

Figure 6.10 shows the data points plotted in the scattering diagram obtained by the PCA. Here, eight-dimensional space made from eight channel outputs was reduced to a two-dimensional plane. PC1 in Fig. 6.10 reflects

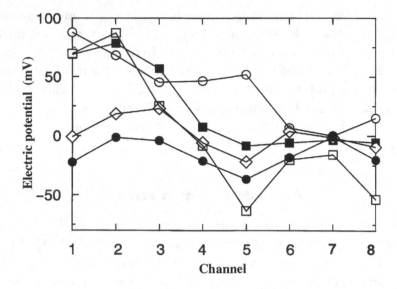

Figure 6.9. Taste sensor response to amino acids:[3] 10 mM L-glutamic acid (○); 10 mM L-tryptophan (■); 100 mM monosodium L-aspartate (□); 100 mM L-value (◇); 100 mM L-alanine (●).

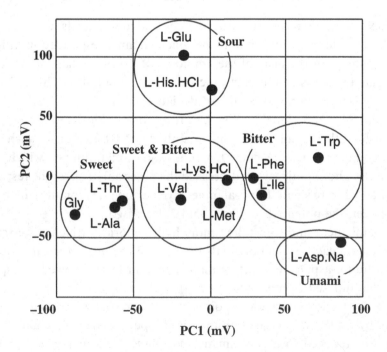

Figure 6.10. PCA of the response patterns for amino acids.[3]

bitterness and sweetness; PC2 reflects sourness and umami taste. Amino acids are classified clearly into five groups by the taste sensor.

The contribution rates were 47.0, 34.0 and 13.8% for PC1, PC2 and PC3, respectively. The rate for PC3 is still relatively large; this indicates that amino acids clearly show three or four independent taste qualities, as is known from neurophysiological and psychophysical studies.

The response of the sensor to amino acids was compared with the results of panel tests[26] for the intensities of five taste qualities obtained for each amino acid. Response potentials from the eight membranes were transformed into five basic tastes by multiple linear regression. This expression of five basic tastes reproduced the human taste sensation very well.[3]

Table 6.2 gives data on hydrophobicity of amino acids.[29] The listed values indicate the free energy change Δf calculated from the solubility data of individual amino acids. It was suggested that amino acids are not bitter if $\Delta f < 1300$, while they are bitter if $\Delta f > 11400$ (see also Table 2.2). The positive value of Δf means that solution in water is thermodynamically unfavorable. Therefore, amino acids with larger Δf values are barely soluble in water. In other words, amino acids with larger Δf have hydrophobic characteristics, whereas those with smaller Δf are hydrophilic. Therefore, we call Δf "hydrophobicity" hereafter.

As a next step, the relationship between the sensor outputs in Fig. 6.9 and the hydrophobicity of amino acids in Table 6.2 was studied.[23] For this purpose, we used a stepwise method of the type used in multiple regression analysis. The method is as follows. The multiple regression analysis was applied to express the hydrophobicity Δf (objective variable) using eight

Table 6.2. *Hydrophobicity of amino acids*

Amino acid	Hydrophobicity Δf
L-Alanine (Ala)	730
Monosodium L-aspartate (Asp. Na)	540
L-Glutamic acid (Glu)	550
Glycine (Gly)	0
L-Isoleucine (Ile)	2970
L-Lysine monohydrochloride (Lys. HCl)	1500
L-Methionine (Met)	1300
L-Phenylalanine (Phe)	2650
L-Threonine (Thr)	440
L-Tryptophan (Trp)	3000
L-Valine (Val)	1690

explanatory variables of response electric potentials of all eight channels. The multiple regression coefficient ($\hat{r}$: adjusted r) was calculated to give the quantitative agreement between objective and explanatory variables. Next, the number of explanatory variables was reduced to seven by omitting one of them, and then $\hat{r}$ was calculated for the resulting eight cases in the same way. The same procedure was carried out by reducing the number of explanatory variables step by step. During this procedure, one set of explanatory variables was adopted when $\hat{r}$ was a maximum, because the objective variable was considered to be expressed most quantitatively in this case.

The result is shown in Fig. 6.11, where we can see a good correlation between the sensor outputs and the hydrophobicity. The adjusted correlation coefficient $\hat{r}$ was 0.977, which seemed to be sufficiently high for explaining the objective variable, i.e., the hydrophobicity of amino acids. The selected set of five channels comprised ch. 1 (DA), ch. 2 (OA), ch. 3 (DOP), ch. 5 (DOP:TOMA 5:5) and ch. 7 (TOMA). This result is satisfactory; in Fig. 6.10, chs. 1 and 3 contributed significantly to PC1, in which the bitter taste of amino acids was distinguished from the sweet taste, whereas chs. 6 and 8 hardly contributed to it.

When the hydrophobicity was expressed by only two channel outputs as a trial, the resulting $\hat{r}$ became 0.723, which also seemed high, using DOP and

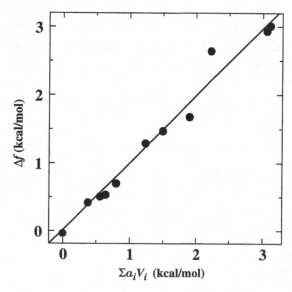

Figure 6.11. Relationship between hydrophobicity of amino acids and sensor outputs.[23] The abscissa represents the multiregression linear equation with V_i denoting the output of channel i ($= 1, 2, 3, 5, 7$) and a_i the multiregression coefficient; the ordinate represents the data on hydrophobicity of amino acids in Table 6.2.

TOMA membranes. The result is very interesting because these two membranes are negatively and positively charged, respectively. The response characteristics are very different between these two (see Fig. 6.5). The hydrophobicity, which may be related to bitter taste, can be quantified using two lipid/polymer membranes with different characteristics.

The reason why the taste of amino acids was discriminated by the taste sensor is partly owing to the strong correlation of sensor outputs with the hydrophobicity of amino acids, which may be a global property of molecules. Grouping of the tastes of amino acids into bitter, sweet, sour and umami taste may reflect this kind of property, which is determined from the hydrophobicity and hydrophilicity of molecules, whereas a molecular recognition by specific proteins of the biological membrane may also have an effect on taste reception. During initial reception of chemical substances producing bitterness, sourness and saltiness, the global selectivity, which originates from hydrophobic and hydrophilic interactions between the biological membrane and chemical substances, is considered to be very important.

6.4.2 Measurement of bitter taste and production of mixed taste

Next, we compared the response patterns for bitter-tasting amino acids such as L-tryptophan with that of quinine, which is a typical bitter substance.[24,25] Of course, these two chemicals have very different chemical structures; nevertheless, we sense the same bitter taste quality. Is it possible to show the same bitter taste using the taste sensor with lipid/polymer membranes? Figure 6.12 shows the comparison of response patterns of three amino acids (L-alanine, L-tryptophan, L-phenylalanine) with those of a bitter substance (quinine), sour substance (HCl) and umami taste substance (MSG). The response patterns were normalized using the formula:

$$v_i = \frac{V_i}{\sqrt{\sum_{i=1}^{8} |V_i|^2}}, \qquad (6.1)$$

where V_i denotes the response electric potential of channel i. To make it easy to see the pattern, the response electric potentials of the positively charged membranes (chs. 6–8 in Table 6.1) and the noncharged membrane (ch. 5) were reversed, because they usually have a sign opposite to those of the negatively charged membranes (chs. 1–4).

Figure 6.12 includes three normalized patterns for one chemical substance with three different concentrations. By the normalization procedure of eq. (6.1), the three patterns of each chemical substance agree with each other; that is, the pattern is independent of the concentration. This fact implies that

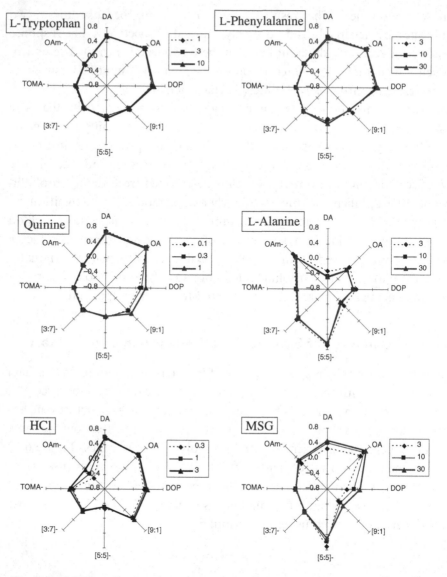

Figure 6.12. Normalized patterns for L-tryptophan, L-phenalanine, L-alanine, quinine, HCl and MSG.[24,25] The three numerical figures closed by a square attached to each pattern imply the concentration (mM). (With permission from Nagamori and Toko.[24])

each chemical substance has an original pattern characteristic of each taste quality. The patterns of amino acids such as L-tryptophan and L-phenylalanine are similar to that of quinine. However, they differ from those of other taste substances such as L-alanine, HCl and MSG. This impression can be strengthened by looking at Fig. 6.7, where these three bitter substances are located close together.

The patterns for L-tryptophan and L-phenylalanine jut out to the upper-right direction (i.e., at chs. 1–3), whereas the pattern for L-alanine shows the opposite tendency: its pattern bulges out to the lower-left direction. The bitter chemical quinine shows a bulge to the upper-right direction in a similar way to L-tryptophan.

The correlation coefficients are shown in Table 6.3. Although L-tryptophan has low correlations with salty (NaCl), sweet (sucrose) and umami (MSG) taste substances, it has a high correlation with a bitter substance, quinine. This result indicates that L-tryptophan shows the same taste as quinine; i.e., L-tryptophan tastes bitter.

Comparison of the original response pattern of L-tryptophan with that of quinine makes it possible to estimate the bitter strength of L-tryptophan in terms of the quinine concentration (see also Section 6.9). As a result, it was concluded that 10 mM L-tryptophan has the same bitter strength as 0.02 mM quinine. To confirm it, we performed the sensory evaluation using our tongue. The result supported the estimation using the taste sensor: humans felt the same bitter strength as 10 mM L-tryptophan by taking 0.02–0.03 mM quinine.

This result implies that the taste sensor measures the taste in itself, as humans do, irrespective of the difference in chemical structures of amino acids and alkaloids such as quinine.

Let us mention a more impressive result concerning the production of taste. L-Methionine and L-valine shows bitter and sweet tastes simultaneously, as listed in Table 2.2. We tried to produce the complicated mixed taste of these amino acids using the combination of a sweet amino acid, L-alanine, and a bitter amino acid, L-tryptophan.

Figure 6.13 shows the normalized patterns for L-methionine and the mixed solution, which contains 100 mM L-alanine and L-tryptophan at 1, 3 and 10 mM. Three patterns for L-methionine at three different concentrations agree fairly well with each other. In addition, they differ from those for either L-alanine or L-tryptophan alone, as shown in Fig. 6.12 (Table 6.4). For example, 10 mM L-methionine has the low correlations of 0.22 and 0.57 with 10 mM L-alanine and 1 mM L-tryptophan, respectively. This implies

Table 6.3. *Correlation coefficients between L-tryptophan and five basic tastes*

	Quinine 0.3 mM	NaCl 30 mM	HCl 3 mM	Sucrose 100 mM	MSG 10 mM
L-Tryptophan 10 mM	0.903	0.276	0.763	0.515	0.408

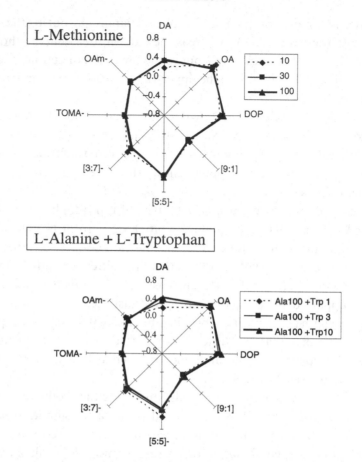

Figure 6.13. Normalized patterns for L-methionine and the mixed solution, which contains 100 mM L-alanine and L-tryptophan at three different concentrations (1–3, 10 mM).[24,25] (With permission from Nagamori and Toko.[24])

that L-methionine has an original taste quality that is different from pure sweet or bitter taste.

The patterns for the mixed solutions composed of 100 mM L-alanine and L-tryptophan of three different concentrations are not the same. Increasing the L-tryptophan concentration tends to cause a bulge to the upper-right direction. Since the pattern for L-tryptophan alone has a characteristic bulge of this type, it implies that the taste becomes nearer that of L-tryptophan, as expected.

Table 6.4 summarizes the correlation coefficients among L-methionine, L-alanine, L-tryptophan and the above three mixed solutions. The response patterns for L-alanine have negative correlations with those for L-tryptophan, which agrees with their very different taste qualities. We can see that the highest correlation is realized between one of the mixed solutions and

Table 6.4. Correlation coefficients for L-methionine, L-alanine, L-tryptophan and the three mixed solutions

		Methionine (mM)			Mixed[a] (Trp, mM)			Alanine (mM)			Tryptophan (mM)		
		10	30	100	1	3	10	10	30	100	1	3	10
Methionine (mM)	10		0.96	0.93	0.93	0.96	0.93	0.22	0.26	0.34	0.57	0.60	0.61
	30	0.96		0.99	0.90	0.97	0.99	0.10	0.13	0.23	0.66	0.70	0.73
	100	0.93	0.99		0.89	0.96	0.98	0.11	0.14	0.24	0.64	0.68	0.72
Mixed[a] (Trp, mM)	1	0.93	0.90	0.89		0.95	0.88	0.48	0.52	0.60	0.33	0.38	0.40
	3	0.96	0.97	0.96	0.95		0.98	0.20	0.24	0.34	0.59	0.64	0.65
	10	0.93	0.99	0.98	0.88	0.98		0.02	0.06	0.16	0.72	0.76	0.78
Alanine (mM)	10	0.22	0.10	0.11	0.48	0.20	0.02		1.00	0.99	−0.66	−0.62	−0.61
	30	0.26	0.13	0.14	0.52	0.24	0.06	1.00		0.99	−0.63	−0.59	−0.58
	100	0.34	0.23	0.24	0.60	0.34	0.16	0.99	0.99		−0.56	−0.51	−0.49
Tryptophan (mM)	1	0.57	0.66	0.64	0.33	0.59	0.72	−0.66	−0.63	−0.56		1.00	0.99
	3	0.60	0.70	0.68	0.38	0.64	0.76	−0.62	−0.59	−0.51	1.00		1.00
	10	0.61	0.73	0.72	0.40	0.65	0.78	−0.61	−0.58	−0.49	0.99	1.00	

[a] Mixed: 100 mM L-alanine + L-tryptophan.

L-methionine. Increasing L-tryptophan concentration in the mixed solution leads to both a decrease of correlation with L-alanine and an increase of correlation with L-tryptophan, a reasonable change. The highest correlation 0.99 is realized for 30 mM L-methionine using the mixed solution 100 mM L-alanine and 10 mM L-tryptophan. The sensory tests by humans agreed with this result. A similar result was obtained for L-valine, which elicits bitter and sweet tastes simultaneously. The sensor output showed the highest correlation between 30 mM L-valine and 100 mM L-alanine plus 3 mM L-tryptophan, as felt by humans.

Dipeptides elicit various taste qualities depending on the amino acids involved. For bitter dipeptides such as Gly–Leu, Gly–Phe and Leu–Gly, the taste sensor showed similar response patterns to those for L-tryptophan (T. Nagamori and K. Toko, unpublished data). In addition, the patterns characteristic of sourness, which is elicited by amino acids such as L-glutamic acid and L-histidine monohydrochloride, were obtained for dipeptides such as Gly–Asp, Ser–Glu, Ala–Glu and Gly–Gly, which taste sour. Dipeptides such as Ala–Gly and Gly–Glu have little or no taste. Small response patterns were obtained for these dipeptides, as expected.

As explained in Chapter 2, it is believed that receptors of amino acids differ physiologically from those of alkaloids such as quinine. As found here, however, the bitterness of amino acids can be expressed by the response patterns of the taste sensor, which uses the lipid membranes, in a similar way to the bitterness of quinine. The mixed taste of amino acids can also be produced using the taste sensor. This result suggests that it may be the lipid (hydrophobic) part of biological membranes that forms the receptor for bitter taste.

6.5 Expression of taste by basic taste qualities

6.5.1 Production of the taste of commercial drinks with basic taste substances

An artificial taste solution was made by combining basic taste substances using the electric potential output patterns to create a solution that was similar to some commercial drinks.[30] As the basic taste substances, HCl, NaCl, sucrose and quinine were chosen for sourness, saltiness, sweetness and bitterness, respectively.

Four different concentrations were prepared for each of these substances: 1, 3, 10, 30 mM for HCl; 30, 100, 300, 1000 mM for NaCl and sucrose; and 0.03, 0.1, 0.3, 1 mM for quinine. The lowest concentrations correspond nearly to the thresholds of detection in humans. We prepared 4^4 (256) mixed solutions with different compositions by combining these four basic solutions.

The 256 mixed solutions were measured with the multichannel taste sensor to give data on the output electric potential patterns for each solution. While the data on each channel output V_i ($i = 1$–8) were dispersed discretely in the five-dimensional space constructed from V_i and four different taste qualities, we approximated V_i by a quadratic function of the concentrations. As a result, eight quadratic functions were obtained for V_i.

As test drinks, two commercial drinks of different brands were chosen. Figure 6.14 shows the electric potential patterns of these two drinks. While they are not very different, the discrimination is easy because of standard deviations of 2 mV. We attempted to fit the patterns from the artificial solutions to the output electric potential pattern of one of the commercial drinks (let us call it "drink A" for convenience) by minimizing the distance between the output electric potentials for the mixed solutions (expressed by the quadratic function) and drink A. The distance is introduced by

$$L = \sum_{i=1}^{8} (V_{mi} - V_{Ai})^2,$$
(6.2)

where V_{mi} and V_{Ai} are the output potentials for the mixed solutions and drink A, respectively, with i denoting the output from ch. i. As a consequence, we

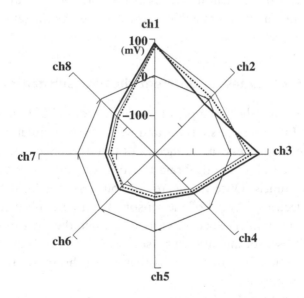

Figure 6.14. Comparison of commercial drinks with an artificial mixed solution.[30] Drink A, ·····; another drink, ▬▬; mixed solution tasting to humans very similar to drink A, ▬▬.

obtained one mixed solution whose output pattern was nearest to that of drink A.

The best combination of the concentrations for basic taste substances was 2 mM HCl, 50 mM NaCl, 0.2 mM quinine and 100 mM sucrose. As can be seen from Fig. 6.14, the pattern for the mixed solution is closer to that of drink A than to another commercial drink. The sensory evaluation by humans was also made. The two commercial drinks and the mixed solution were tasted, and their tastes were compared. The sensory tests by humans showed that this mixed solution produced almost the same taste as drink A.

It is very important to note that this method automatically contains the interactions between taste substances. If many measurements of various mixed solutions are made, comparison of output patterns between the test solution and the mixed solutions can be easily made by adequate algorithms, such as in neural networks.

The τ scale is effective to express the taste strength; it represents the subjective strength of a taste solution.[31,32] The concentration of each taste substance can be transformed into the taste strength. The above mixed solution is thus composed of 4.04 sourness, 2.03 saltiness, 5.01 bitterness and 2.24 sweetness in terms of the τ scale, and drink A has the above taste qualities and intensities.

The taste of every drink can be quantified using the multichannel taste sensor. In other words, quantitative measurements of taste are possible by the sensor that are more accurate than those achieved by the sense of humans.

6.5.2 Sourness of different chemical substances

Let us here show another example of quantification of taste made with the taste sensor.[33] Intensities of sourness and saltiness were quantified by introducing "degree of sourness" as a function of tartaric acid concentration. The degree was expressed as an algebraic function of outputs of two channels among eight channels. Once the degree of sourness is defined by means of the response electric potential of the sensor, it can be applied to other sour substances. Correspondence between this sensor and the sense of human taste was studied. The sensor and human sensory evaluation of HCl, tartaric acid, citric acid and lactic acid matched each other; hence this sensor reproduces the human taste sensation.

When tartaric acid and NaCl occur together, they enhance each other.[7] This phenomenon was reproduced using the degree of sourness. Thus, the taste sensor expresses the taste interactions experienced by humans.

The sensor could detect minute differences of taste between NaCl, KCl, KBr, NaBr, NH$_4$Cl, LiCl and KI.[33] The response patterns are different, as can be seen from Fig. 6.7 for NaCl, KCl and KBr. It implies such taste substances as KBr, NaBr and KI do not show the pure saltiness elicited by NaCl. Therefore, the sensor can detect large differences between five taste qualities and, furthermore, can distinguish between these small differences in a single taste quality.

6.6 Mechanism of response

A theory to explain the mechanism of response involving both electrostatic and hydrophobic interactions can be developed and used to explain the experimental data in a quantitative manner.[34,35] Variations in the properties of the chemical substances producing taste affect their interaction with lipid/polymer membranes and cause the difference in response electric patterns, as shown in Fig. 6.5.

6.6.1 Responses of the negatively charged membrane to NaCl and quinine

The characteristic upward response of the negatively charged lipid/polymer membrane to NaCl in chs. 1–3 of Fig. 6.5 is not caused by the usual screening effect but by the change in its surface charge density owing to dissociation of protons from the membrane. In most conventional polymer membranes such as colloidal membranes[36] and lipid-adsorbed membrane filters (see Chapter 3), the electric potential tends to saturate at high NaCl concentrations. This is a consequence of the screening effect by ions of opposite charge, which weakens the effect of the electric charge of the membrane. As a result, the response to ions becomes saturated at high ion concentrations.

A schematic of the charged membrane system is shown in Fig. 6.15. We consider the DOP membrane, which is used in ch. 3 in Table 6.1, as a typical negatively charged membrane. The charged membrane separates two KCl solutions I and II. Aqueous solution II refers to the external solution, which contains a taste substance. The membrane potential usually consists of the surface electric potential formed in the aqueous phase touching the membrane and the diffusion potential within the membrane, as shown in Fig. 6.15. In this theoretical model, therefore, the membrane potential is obtained from the sum of the surface potential and the diffusion potential, which are calculated individually.

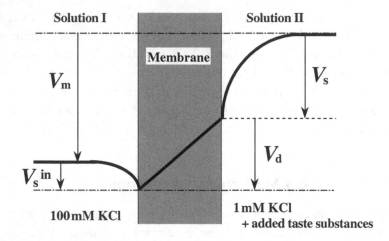

Figure 6.15. The membrane electric potential in a charged membrane system. Solutions I and II are internal and external solutions, respectively. A taste substance was added to the external solution, from which the membrane potential was measured as the origin.

First, we describe the change in the surface potential with the ion concentration in the bulk solution. For this purpose, the change in the surface charge density caused by the hydrophilic groups of the lipid of the membrane that are in contact with the aqueous phase must also be taken into account. It implies that the theory treats the situation where the surface electric potential is changed with H^+ dissociation from lipid molecules, which causes the change in electric charge density, as occurs in some colloidal systems.[37,38] This situation is illustrated in Fig. 6.16.

The change in Gibbs free energy per lipid molecule (dG) with the above dissociation process in the lipid membrane is given from standard thermodynamics by:[37,39]

$$dG = k_B T \ln\left[\frac{K}{[H^+]}\frac{\theta}{1-\theta}\right]d\theta + A V_s d\sigma. \tag{6.3}$$

In eq. (6.3), θ is the degree of H^+ binding to a lipid molecule at a hydrophilic group, K is the dissociation constant, $[H^+]$ the proton concentration in the bulk solution, k_B the Boltzmann's constant, T the absolute temperature, σ the surface charge density, A the occupied molecular surface area per lipid molecule and V_s is the surface electric potential of the membrane. We obtain an expression for the free energy in the charged membrane system as

$$G = \int_0^\theta k_B T \ln\left[\frac{K}{[H^+]}\frac{\theta}{1-\theta}\right]d\theta + A\int_0^\sigma V_s d\sigma. \tag{6.4}$$

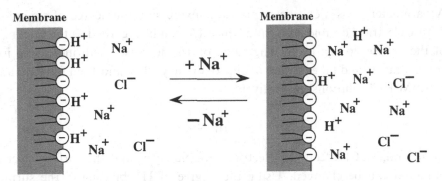

Figure 6.16. The situation by which the surface electric charge density of the membrane is changed by H$^+$ dissociation from lipid molecules with changing ion concentration in the bulk solution.

For simplicity, the membrane surface is regarded as a plane surface with the uniform charge density σ.

The charge density σ at the membrane surface is determined using the Gouy–Chapman theory of the electrical double layer; that is, the ion distribution near the membrane surface and the surface charge density are calculated by solving the Poisson–Boltzmann equation.[37-39] For 1 : 1 electrolyte we get the following Poisson–Boltzmann equation.

$$\frac{d^2}{dx^2} V(x) = \frac{8\pi e}{\varepsilon} c \sinh\left(\frac{eV}{k_B T}\right), \tag{6.5}$$

where x is the spatial coordinate inside the aqueous solution normal to the concerned membrane surface, ε the dielectric constant, e the elementary charge and c denotes the ion concentration in the bulk solution.

The boundary conditions for $V(x)$ are given by

$$\frac{dV}{dx} = -\frac{4\pi}{\varepsilon}\sigma \qquad \text{at } x = 0, \tag{6.6a}$$

$$V = \frac{dV}{dx} = 0 \qquad \text{at } x = \infty. \tag{6.6b}$$

With the aid of these boundary conditions, eq. (6.5) can be integrated and eq. (6.6a) is reduced to

$$\sigma = \kappa' \sinh\left(\frac{eV_s}{2k_B T}\right); \tag{6.7}$$

$$\kappa' = \frac{\varepsilon}{2\pi} \frac{k_B T}{e} \kappa, \qquad \kappa = \sqrt{\frac{8\pi c e^2}{\varepsilon k_B T}}. \tag{6.8}$$

A parameter, κ, is a characteristic value expressing the degree of spread of diffuse electrical double layer, and hence $1/\kappa$ can be regarded as the thickness of the diffuse double layer. The value of $1/\kappa$ decreases with increasing ion concentration and takes values of approximately 10, 3, and 1 nm for 1, 10 and 100 mM KCl solutions, respectively.

Case of NaCl

In general, NaCl affects the electrical double layer, and the surface electric potential can be changed. Using the degree of H^+ binding θ, the surface charge density σ is also expressed as the following equation:

$$\sigma = -\frac{e}{A}(1 - \theta). \tag{6.9}$$

If we eliminate σ from eq. (6.7) and eq. (6.9), the following equation is obtained:

$$\kappa' \sinh\left(\frac{eV_s}{2k_B T}\right) = -\frac{e}{A}(1 - \theta). \tag{6.10}$$

By minimizing eq. (6.4) with respect to θ by taking account of eq. (6.9), we get the following equation.

$$\frac{\theta}{1 - \theta} = \frac{[H^+]}{K} \exp\left(\frac{-eV_s}{k_B T}\right). \tag{6.11}$$

In the two equations (6.10) and (6.11), two variables θ and V_s are unknown. Therefore, the surface potential V_s of the membrane can be calculated as a function of the ion concentration c.

Case of quinine

Since quinine is a hydrophobic molecule, it may be natural to consider that quinine ions are bound to the hydrophobic part of a membrane. Therefore, the surface charge density σ of the membrane is changed by this binding effect; proton binding to hydrophilic site and the quinine binding should be separately measured and yet these two effects are not independent. The expression for σ is then:

$$\sigma = -\frac{e}{A}(1 - \theta) + \frac{e}{A}\theta_q, \tag{6.12}$$

where θ_q is the degree of binding of quinine ions. This is dependent on the quinine concentration near the membrane surface and the electric charge condition of the membrane. Therefore, the expression for θ_q is:

$$\theta_q = a(1 - \theta)^2 c_q \exp\left(-\frac{eV_s}{k_B T}\right), \tag{6.13}$$

where c_q is the bulk quinine concentration and a is a numerical parameter. The factor $(1 - \theta)^2$ assumes that more quinine ions are bound to the membrane because of nonoccupied sites at the surface at the first step of binding.

The term σ can be eliminated from eqs. (6.7) and (6.12), and an equation similar to eq. (6.10) can be obtained. In the present case, too, ions such as H^+ (of H_2O) and K^+ (of 1 mM KCl) are contained in the aqueous medium; therefore, eq. (6.11) holds.

Next, let us consider the diffusion potential V_d within the membrane. In charged systems shown in Fig. 6.15, K^+, Cl^- and T^+ usually diffuse into the membrane, where T^+ denotes counterions from the electrolyte of taste substances in the external solution. As a result, the diffusion potential is generated by the difference in the mobility of cations and anions. The diffusion potential V_d within the membrane is calculated from the following Goldman–Katz equation:[40,41]

$$V_d = \frac{k_B T}{e} \ln \left[\frac{\mu_K c_K^{ex} + \mu_{Cl} c_{Cl}^{in} + \mu_T c_T^{ex}}{\mu_K c_K^{in} + \mu_{Cl} c_{Cl}^{ex}} \right], \qquad (6.14)$$

where μ_i denotes the product of the mobility of ion i (K^+, Cl^-, T^+) within the membrane and the partition ratio of ion i to the two phases, which are the membrane and the external solution, and c_i the i ion concentration near the membrane surface. The superscripts ex and in represent the membrane surface on the side of external solution and that on the side of another solution across the membrane, respectively.

The membrane potential defined in Fig. 6.15 is explained as

$$V_m = V_s + V_d - V_s^{in}, \qquad (6.15)$$

where V_s^{in} is the surface electric potential formed in the aqueous phase of another side across the membrane.

Theoretical results are compared with the observed response potential of the DOP membrane (ch. 3) to NaCl and quinine in Fig. 6.17, where the solid curves represent the theoretical results.[34] The theoretical data and the observed data on response potentials of the membranes agree quantitatively for both NaCl and quinine. The theoretical curve of response potential is taken relative to a standard potential calculated at pH 5.8, and 300 K, with c as 1 mM for 1 mM KCl solution without taste substances. The theoretical value of the standard potential of the lipid/polymer membrane is about -55 mV, while the observed absolute value was about -50 mV. The parameter values were chosen to explain the experimental results in the best fit as a whole within their reasonable ranges: $K = 10^{-4}$ M, $A = 120$ Å^2 (1.2 nm^2), $\mu_K = \mu_{Cl} = \mu_T = 0$ and $a = 150$. The occupied molecular surface

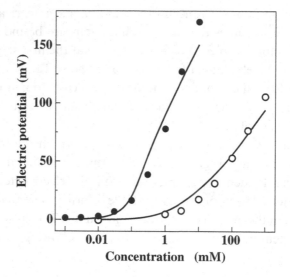

Figure 6.17. Response potential of the lipid/polymer membrane to NaCl (○) and quinine
(●).[34] The solid lines represent the theoretical results.

area A was estimated from the volume of the membrane and the quantity of
used lipid DOP. We assumed that A was constant even when taste substances
were adsorbed in the membrane as a first approximation.

In the responses of the membrane to NaCl and quinine, the surface potential contributes to most of the response potential. The diffusion potential is
hardly generated because lipids packed densely in the membrane interfere
with ion permeation. It implies that the lipid/polymer membrane made
of DOP is such that ions can hardly permeate through the membrane,
whereas the inside of membrane is electrically connected to the outer surface
because of the hydrophilic part of lipids. This is very reasonable because
the membrane electric resistance is as high as several megaohms per square
centimeter, as mentioned in the next chapter. In this high-resistance state, ion
permeability cannot be expected.[42–44] Among the lipid/polymer membranes
listed in Table 6.1, all except the TOMA membrane have such a high electric
resistance. Therefore, the membrane potentials of the negatively charged
membranes originate mainly from the surface potential, whereas both the
surface and diffusion potentials cannot be neglected in the TOMA membrane.

In usual cases of fully charged membranes, Na^+ scarcely changes the net
electric charge density of a negatively charged membrane but affects the
surface electric potential (i.e., the electric screening effect), which tends to
be flat at high NaCl concentrations, as seen in Fig. 3.23 (p. 67). Let us now
consider the response characteristics for NaCl in Fig. 6.17, where electric

potential changes of the lipid/polymer membrane are not saturated at higher NaCl concentrations.

Figure 6.18 shows the calculated results of degrees of H^+ dissociation from a lipid in the membrane. In the lipid/polymer membrane, H^+ is scarcely dissociated from a lipid at low NaCl concentrations, as indicated by the low degree of H^+ dissociation of 0.21 in Fig. 6.18. It is because the occupied molecular surface area of used lipid is small owing to the method of membrane preparation; lipid is packed densely in the membrane. This dense packing inhibits the charging process (i.e., dissociation of H^+ from the hydrophilic group of the lipid), which causes strong electric repulsion between charged molecules. The dissociation of H^+ is accelerated by the addition of NaCl. This implies an increase in the magnitude of the surface electric charge density, as seen in Fig. 6.19 where the calculated surface charge density is shown as a function of NaCl or quinine concentration; in this case, the membrane becomes charged more negatively. Increasing NaCl concentration weakens the electric repulsion between lipids, and hence H^+ becomes dissociated.

Therefore, it is reasonable to believe that the lipid/polymer membrane is not electrically charged to a great extent at lower ionic strength and becomes more negatively charged through the accelerating dissociation of H^+ with increasing NaCl concentration. As a result, the rate of change of the electric potential increases with increasing NaCl concentration, since the membrane becomes more negatively charged and the sensitivity to cations such as Na^+ is increased.

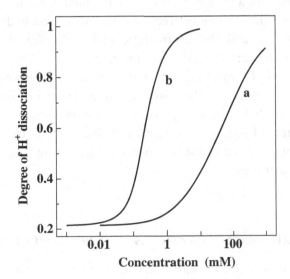

Figure 6.18. The degrees of H^+ binding to the membrane for taste substances in (a) NaCl and (b) quinine.[34]

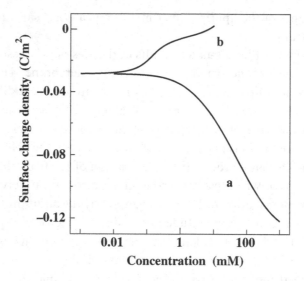

Figure 6.19. The surface charge densities of the membrane for taste substances in (a) NaCl and (b) quinine.[34]

For quinine, the membrane potential of the lipid/polymer membrane changes more than that for NaCl, as seen from Fig. 6.17. Quinine hydrochloride is a hydrophobic molecule with a hydrophilic positively charged portion. Hence quinine ions bind to the hydrophobic part of negatively charged membranes and lead to large changes in the membrane potential. The degree of binding of a taste substance with the membrane may be dependent on two factors; the balance between hydrophilicity and hydrophobicity of the taste substance and the electrostatic and hydrophobic interaction between the membrane and the taste substance.

Figure 6.20 shows the calculated degree of binding of quinine ions to the lipid/polymer membrane. The dissociation of H^+ from lipids in the membrane is accelerated with increasing quinine concentration, as seen in Fig. 6.18; the degree of binding of quinine ions increases at the same time. The binding is brought about by the hydrophobic interaction between quinine and the lipid/polymer membrane.

6.6.2 Hybrid membranes composed of two lipid species

Biological membranes are highly stable and have complex functions. As these membranes contain a number of lipid species, it became important to examine the basic characteristics of artificial hybrid membranes.

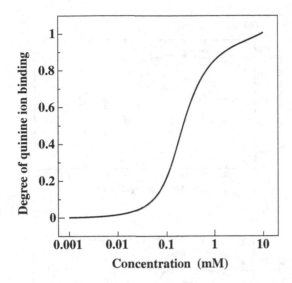

Figure 6.20. The degree of quinine ion binding to the membrane as a function of quinine concentration.[34]

Electrical characteristics of hybrid membranes composed of two lipid species were studied[35] where one lipid species is positively charged in aqueous solution and the other is negatively charged.

The hybrid membranes comprised two lipid species, DOP and OAm, at molar ratios of 0 to 100% in 10% steps. In aqueous solution at neutral pH, DOP is negatively charged by dissociation of the phosphoric acid group, while OAm is positively charged by ionization of the ammonium group. Two lipids were mixed with the desired mixing ratio in a test tube with a polymer (800 mg PVC) and a plasticizer (1.0 ml DOPP) and dissolved in 18 ml tetrahydrofuran. The mixing ratio implies the molar ratio of OAm to the total quantity of lipids (200 µmol) including DOP.

Figure 6.21(*a*) shows the response potential patterns of the hybrid membranes to HCl, which produces sourness. The responses of the hybrid membranes with a mixing ratio less than 50% were positive, whereas those of over 60% were negative. The response of each hybrid membrane varied in magnitude with the mixing ratio. This suggests that each hybrid membrane was different in electric charge condition in an aqueous medium, and that the electric charge of the hybrid membrane inverted at approximately the 50% mixing ratio.

The response potential of the negatively charged membrane to HCl became larger as the mixing ratio increased from 0 to 50%; furthermore the change appeared at lower HCl concentrations. This implies that the hybrid

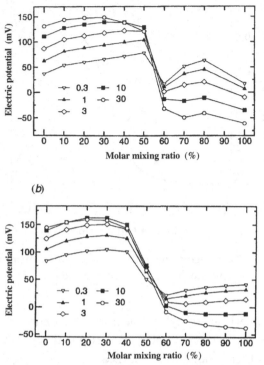

Figure 6.21. Changes in the membrane electric potential with HCl (mM) as found by experiment (*a*) or calculated (*b*).[35] The molar mixing ratio is the ratio of OAm in molar unit to the total lipid.

membranes have a higher sensitivity to HCl than the membrane made of a single lipid, DOP. Moreover, it is noticeable that the upward response to HCl appears in the membranes of ratios 60–100% and then the downward response occurs at higher HCl concentrations.

These responses can be explained quantitatively. The membrane potential of the hybrid membrane is constructed from the diffusion potential within the membrane and the surface electric potential. As shown in Section 6.6.1, the surface electric potential can be considered to contribute mainly to the membrane potential. Therefore, we can describe the change in the surface potential with the HCl concentration by neglecting the diffusion potential.

For hydrophilic groups of each lipid, the following ionization equilibria hold at the hybrid membrane surface.

$$(C_8H_{17})_2POOH: R_2 - POO^- + H^+ \rightleftharpoons R_2 - POOH,$$
$$C_{17}H_{33}CH_2NH_2: R' - NH_2 + H^+ \rightleftharpoons R' - NH_3^+,$$

(6.16)

where R and R$'$ imply hydrophobic hydrocarbon chains.

We obtain an expression for the free energy in the system of the hybrid membrane of DOP and OAm with the ratio $(1 - \alpha): \alpha$ as

$$
G = (1 - \alpha) \int_0^{\theta_C} k_B T \ln \left[\frac{K_C}{[H^+]} \frac{\theta_C}{1 - \theta_C} \right] d\theta_C
$$

$$
+ \alpha \int_0^{\theta_N} k_B T \ln \left[\frac{K_N}{[H^+]} \frac{\theta_N}{1 - \theta_N} \right] d\theta_N + A \int_0^{\sigma} V_s d\sigma, \qquad (6.17)
$$

where θ_C and θ_N are the degrees of H^+ binding to a lipid molecule at hydrophilic groups of DOP and OAm, respectively, K_C and K_N describing the dissociation constants. For simplicity, we assumed the occupied molecular surface area of DOP is equal to that of OAm (put to A).

The surface charge density σ is expressed using the degrees of H^+ binding θ_C and θ_N:

$$
\sigma = (1 - \alpha) \left[-\frac{e}{A} (1 - \theta_C) \right] + \alpha \frac{e}{A} \theta_N. \qquad (6.18)
$$

If we eliminate σ from eqs. (6.7) and (6.18), the following equation is obtained.

$$
\kappa' \sinh \left(\frac{eV_s}{2k_B T} \right) = (1 - \alpha) \left[-\frac{e}{A} (1 - \theta_C) \right] + \alpha \frac{e}{A} \theta_N. \qquad (6.19)
$$

By minimizing eq. (6.17) with respect to θ_C and θ_N by taking account of eq. (6.18), we get the following equations.

$$
\frac{\theta_C}{1 - \theta_C} = \frac{[H^+]}{K_C} \exp \left(-\frac{eV_s}{k_B T} \right), \qquad (6.20a)
$$

$$
\frac{\theta_N}{1 - \theta_N} = \frac{[H^+]}{K_N} \exp \left(-\frac{eV_s}{k_B T} \right). \qquad (6.20b)
$$

In the three equations (6.19), (6.20a) and (6.20b), three valuables θ_C, θ_N and V_s are unknown. Therefore, the surface potential V_s of the hybrid membrane can be calculated for comparison with the experimental data in Fig. 6.21(a) as a function of the HCl concentration for the molar mixing ratio α of lipid species.

Figure 6.21(b) shows the theoretical results of the response patterns. We can see fairly good quantitative agreements with the observed data. The response to HCl was larger, for example, in the membrane with a mixing ratio of 40% than in the single-lipid membrane of DOP; i.e., the sensitivity to HCl increased with the molar mixing ratio from a ratio of 10% to one of 50%. After an initial increase, the electric potential decreased with increasing HCl as the mixing ratio increased from 60%.

The increase in electric potential with HCl in the membranes with ratios of 0–40% was brought about by the binding of H^+ on the membrane surface. By comparison, the electric screening by Cl^- caused the decrease in electric potential with increasing HCl concentration in the positively charged membranes of mixing ratios 60–100%, after H^+ is bound with the ammonium group according to eq. (6.16).

6.7 Measurement of taste of foods

The present sensor could easily discriminate between some commercial drinks such as coffee, beer and ionic drinks, as shown in Fig. 6.22.[30] These three output patterns are definitely different, since the standard deviations were 2 mV at maximum in this experimental condition, where the output for 1 mM KCl was taken as the origin. If the data are accumulated in the computer, any food can be easily discriminated. Furthermore, the taste quality can also be described quantitatively by the method described below. In biological systems, patterns of frequency of nerve excitation may be fed into the brain, and then foods are distinguished and their tastes are recognized using this stored information. In a similar way, the quality control of foods becomes possible using the taste sensor, which has a mechanism of information processing similar to that in biological systems.

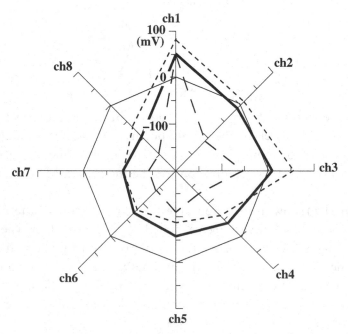

Figure 6.22. Response electric potential patterns for coffee ▬ ▬ ▬ ▬, beer (▬▬▬) and a commercial drink (- - - - -).[30]

6.7.1 Beer

Figure 6.23 shows the response patterns to eight different brands of beer.[45] The patterns were measured relative to a certain beer as a standard. Carbon dioxide in the standard beer was removed by stirring it for one night. The measurement (the first rotation) was started about 3 min after each sample beer, which was just opened, was set at the regular position. Although the difference in electric potential between different brands was a few millivolts or more, each beer was easily distinguished from the other by these patterns because of the high reproducibility and sensitivity of the sensor represented by 0.2 mV standard deviations, which are smaller than the usual condition by one order. This superior ability is a consequence of two aspects of the method.[46] First, a certain standard beer is used as an initial reference for the electric potential pattern although KCl solution was used as the initial reference in Fig. 6.22. The electrodes were immersed in the same standard beer for 4 weeks (preconditioning method). Second, regular repetitive measurements with a period of approximately 20 s were used. This preconditioning with repetitive measurements is very effective when slight differences in taste need to be distinguished.

The reproducibility in short-term measurements, for example one day, is very high, as shown below in standard deviations in each measurement. However, slow drift of the response electric potential occurs for several months. Nevertheless, reliable measurement over a long period becomes possible if a basic standard sample with stable quality is prepared, because the difference of electric potential pattern between the sample and the basic

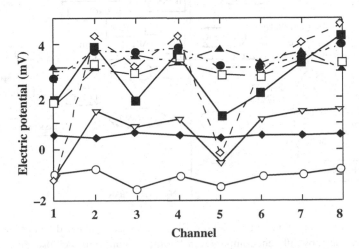

Figure 6.23. Response electric potential patterns for beer.[45]

standard sample is taken as the response electric potential pattern. Of course, the long-term reproducibility also depends on the measured sample. The same membrane can be used continuously for over one year.

The human sense of taste is vague and largely depends on subjective factors of human feelings. Of course, suitable panel methods and statistical examination by humans are fairly reliable. However, daily and real-time tasting of foods is very laborious. The taste sensor can help in this laborious work. In addition, taste can be assessed objectively by comparing the standard index measured by means of the taste sensor with the human sensory evaluation.

The taste of beer can be expressed by four terms: rich, light, sharp and mild.[47] The rich or light taste may be mainly related to the concentration of barley, whereas the sharp or mild taste may arise from the concentrations of alcohol, hops and so on. We tried to express these taste qualities quantitatively by transforming the output pattern of the taste sensor using PCA. Figure 6.24 shows the result. Comparison with the human taste sense implied that PC1 corresponds to rich taste and light taste, and PC2 to sharp taste and mild taste.

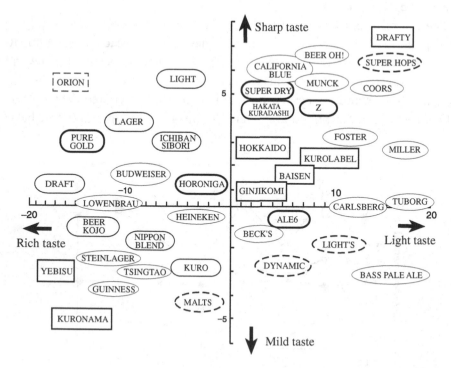

Figure 6.24. Taste map obtained using PCA. Beers enclosed by dashed ellipse and square are those produced by different companies in Japan; the continuous ellipse indicates beer from other countries.

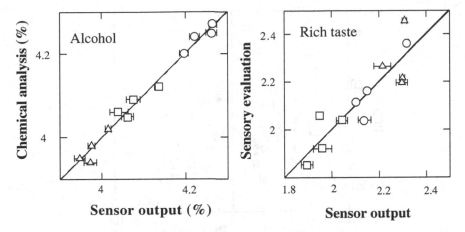

Figure 6.25. Correlations of sensor output with rich taste sensation assessed by humans and the alcohol content of beer.[48] Different symbols indicate different brands of beer.

Figure 6.25 shows correlations of sensor output with rich taste felt by humans and with alcohol concentration.[48] A good linear relationship between the resultant sensor output and the human sensory expression and the physico-chemical quantity of beer can be found. High correlations with other sensory expressions such as bitter and sharp taste were also found, as easily deduced from Fig. 6.24. The taste sensor also showed high correlations with physico-chemical quantities such as pH and the bitter value estimated from iso-α-bitter acid. These results imply the usefulness of the taste sensor for quantifying human sensory expressions and physicochemical quantities in beer.

Humans usually experience saturation of the taste sense after multiple tastings. A similar situation may also occur in the taste sensor; however, repetitive rinsing of the electrode with water (or adequate rinsing fluid) after each measurement enabled reliable, reproducible results to be obtained. As a result, there is no saturation in the response of taste sensor.

6.7.2 Mineral water

Recently, interest in drinking water has grown rapidly. Many people are requesting safe and delicious water. The importance of water has also been recognized as a result of interest in environmental pollution issues. However, no convenient measuring system sufficient for quality evaluation of drinking water is available.

Figure 6.26 shows the result of PCA applied to response patterns for 41 commercial mineral waters,[49,50] which were measured as described by Iiyama

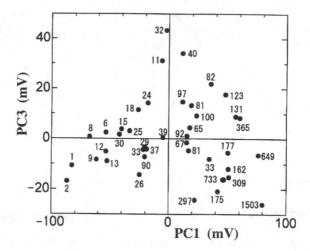

Figure 6.26. Result of PCA applied to taste sensor response patterns for 41 commercial mineral waters.[50] The figures give the hardness of each sample.

and coworkers.[51] We can see that the lower-right plane contains mineral water of a hard type whereas the lower-left plane is occupied by soft water.

The taste of mineral water is quite subtle, and hence it is difficult for humans to discriminate between different brands. However, the taste sensor responded well to mineral water from soft to hard categories. In spite of low concentrations of taste substances in mineral water, the taste sensor can discriminate between different brands because of its high sensitivity to inorganic ions.

Factor loadings of all the channels to PC3 were negative, whereas factor loadings of chs. 1–4 (in Table 6.1) to PC1 were positive and those of chs. 6–8 were negative. Channels 1–4 were made of negatively charged membranes and chs. 6–8 were positively charged; hence PC1 expresses the sum of the effects of cations and anions, while PC3 expresses the simple average of both the effects. The sum and difference of magnitudes of the effects are represented by PC1 and PC3, respectively. If the interaction magnitudes of cations and anions with lipid/polymer membranes are similar to each other, PC3 will become nearly zero. In samples measured here, PC2 had no physical meaning.

Sensory evaluations of seven brands of mineral water were also performed. Four terms expressing the taste were selected: complicated, soft, fresh and sharp. Each term was expressed by five levels, and panelists estimated the level. PCA was carried out on the results of sensory evaluation. PC1 had a contribution rate higher than 93% in all five trials. The transformation matrix showed that PC1 reflects a sharp taste. This implies that the taste of mineral water evaluated by humans has one-dimensional information. In addition, the order of seven brands of mineral water along the PC1 axis differed from trial to trial. That is, good reproducibility of human sensory evaluation was not obtained. This suggests that the expression of the taste of mineral water by human sensory evaluation is difficult.

6.7.3 Other water

Detection of some toxic substances in factory drains is a time consuming process because of the range of substances that need to be analyzed. The taste sensor can be applied to real-time, easy detection of water pollution. After the pollutant is rapidly checked using the taste sensor, the pollution molecule can be analyzed in detail using conventional apparatus for chemical analyses, which are inadequate for real-time detection.

The taste sensor was applied to measurement of contamination of factory drains.[52] Many pollutants such as CN^-, Fe^{3+} and Cu^{2+} could be measured in a few minutes with detection limits lower than those contained in regulations regarding drainage. Cyanide was detected selectively using multiple regression analysis. Detection of heavy metal cations and inorganic anions was tried using several types of ion-selective electrode.[53] Application of multicomponent analysis produced a successful result.

6.7.4 Coffee

Figure 6.27 shows the response electric potentials[54] for 10 brands of coffee from different origins, Brazil (Santos No. 2), Guatemala (SHB), Jamaica

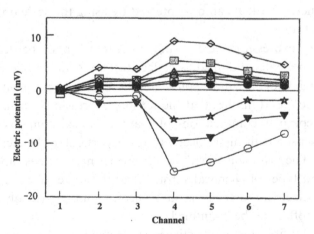

Figure 6.27. Response electric potential patterns for 10 different brands of coffee.[54]
◇ Kenya (AA), ■ Hawaii Kona (Extra Fancy), ▲ Jamaica (Blue Mountain No. 2), △ Tanzania (AA), ◆ Guatemala (SHB), □ Brazil (Santos No. 2), ● Colombia (Excelso), ☆ Indonesia (Mandheling Grade 1), ▼ Indonesia (WIB 1), ○ Indonesia (AP 1).

(Blue Mountain No. 2), Hawaii Kona (Extra Fancy), Kenya (AA), Tanzania (AA), Colombia (Excelso), Indonesia Mandheling (Grade 1), Indonesia WIB 1 and Indonesia AP 1, relatively measured from the electric potential for Salvador (CS), which was used here as the standard coffee. The membranes used were No. SB00 for ch. 1 as a control, No. SB01AAC1 for chs. 2 and 3, No. SB01AAF1 for chs. 4 and 5, and No. SB01AAJ1 for chs. 6 and 7, which were immersed in the standard coffee, Salvador used here, for at least a month. The response electric potentials were taken using this coffee as the origin of electric potential pattern. The membranes of chs. 2 and 3 for coffee measurements were negatively charged, those of chs. 4 and 5 positively charged, and those of chs. 6 and 7 were weakly charged negatively. The measurements were made at 60 °C.

We can see that Kenya shows the largest pattern, while AP 1 shows the smallest pattern. Since the standard deviations of the measuring system were 0.1 mV, these origins of coffee could be easily discriminated.

The basic words characteristic of the taste of coffee are acidity and bitterness. We use the term acidity for coffee instead of sourness, because acidity is always used to express a basic, desirable sharp and pleasing taste produced by organic acids contained in coffee.[55]

Comparisons between the response electric potential and the sensory panel test were made for acidity and bitterness. The correlation coefficients were 0.98 for acidity using ch. 2 and −0.94 for bitterness using ch. 5. This result implies that chs. 2 and 5 can be utilized to quantify acidity and bitterness, respectively. For example, Kenya has strong acidity while Mandheling shows strong bitterness; this fact can be expressed by using the scale obtained from the sensor output.

The sensory expressions acidity and bitterness cannot be deduced from chemical components contained in coffee. In fact, caffeine-free coffee tastes bitter, although caffeine can be considered as a main component to produce bitterness in coffee. It implies that the sensory expressions for taste qualities originate in complex combinations of many chemical components. In this situation, a sensor with high selectivity for a particular chemical substance is not useful. The taste sensor with lipid/polymer membranes can respond to many different types of chemical substance at the same time; i.e., the sensor has a global selectivity to chemical substances. From this reason, acidity and bitterness of coffee can be quantified.

This result is a first step in objective measurement for conventional sensory expressions for coffee production. Desirable blend of several origins of coffee may become possible, for example, by using a method of pattern matching with a data base constructed from measurements using the taste sensor.

6.7.5 Milk

Figure 6.28 shows the time course of responses of five channels to a commercial milk. The response electric potential is the difference between the electric potential for the measured milk and that for another commercial milk as a standard origin. Data sampling was made every second. The response was so fast that it occurs within 1 s, as soon as the electrode is immersed in the sample, and then the response curve of each channel was almost flat for the measuring time 50 s. All the channels showed very slow gradual changes of response electric potentials; this indicates the slow adsorption of chemical components of milk onto the membrane (or desorption from the membrane). Although this slow component of the response occurred, most of the response occurred at the initial time as soon as the membranes were immersed in the sample. A similar situation occurs for other foodstuffs such as beer, coffee and mineral water using preconditioned membranes.

As can be understood from the range of −5 to 5 mV of the response potentials of the channels, most commercial brands of milk have response patterns that differ from others by about 10 mV (the response electric potentials of all the channels are zero for the standard milk in Fig. 6.28).

The sensor was used to assess changes in the taste of milk induced by heat treatment.[56] Some treatments, such as low-temperature long-time or high-temperature short-time, are usually performed on commercial milk for pasteurization or sterilization. These treatments have considerable effects on the taste of milk. Therefore, it is necessary to clarify the effect quantitatively.

Seven heat-treated samples were prepared from the same original milk: 65 °C–0 min, 65 °C–30 min, 80 °C–0 min, 80 °C–15 min, 100 °C–0 min,

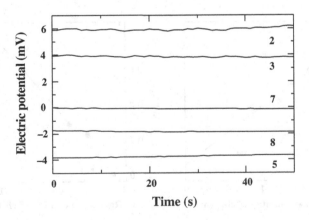

Figure 6.28. Time course of response electric potentials of chs. 2, 3, 5, 7 and 8 (in Table 6.1) in the measurement of milk.

100 °C–15 min and 110 °C–3 s. The temperature was gradually increased to the processing temperature, kept at the temperature during the processing time, and then gradually decreased. Hence, milk with 65 °C–0 min treatment differs from milk with 80 °C–0 min treatment. The precondition for stabilizing the electrochemical property of the membranes was achieved by immersing the electrodes in a standard milk sample (140 °C–2 s heat treatment) for a long period of over a month. The output pattern for the standard milk was used as a reference pattern for all measurements over 6 months.

Since all the samples were of the same origin, the difference in taste among the samples was considered to be slight. It is necessary to detect the electric potentials with high sensitivity and reproducibility. Hence, newly separated standard milk was used as a reference sample and also as the rinsing fluid for the electrodes at every measurement. Each measuring time was set to 60 s, and rinsing of electrodes after each measurement was repeated 10 times for about 20 s.

Three expressions for the taste characteristics of milk, richness, cooked flavor and deliciousness, were evaluated by 15 well-trained panelists for the seven samples.

Whey protein denaturation was also measured by an optical method to obtain the whey protein nitrogen index (WPNI), which was estimated from the absorbance at 420 nm of milk treated with saturated NaCl. A high WPNI value means that a small amount of whey proteins was denatured.[57]

Figure 6.29 shows the relationship between the output from ch. 7 (OAm) and richness and that between ch. 1 (DA) and WPNI, where the correlations are −0.885 and 0.953, respectively. The changes of response potentials of chs. 1 and 7 for different samples are as small as 2 mV, which is much smaller than

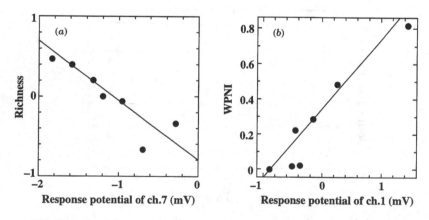

Figure 6.29. Relationship between the output from ch. 7 (OAm membrane) and richness (*a*) and that between ch. 1 (DA membrane) and WPNI of milk (*b*).[56]

the difference of approximately 10 mV found in commercial brands of milk shown in Fig. 6.28. This is logical in that only the heat treatment is different in the present case.

Channels 1 (DA) and 3 (DOP) showed a correlation as high as 0.95 with the degree of protein denaturation caused by the heat treatment of original milk. A sensory test was also carried out; the sensory evaluation did not show a high correlation for the taste sensor with deliciousness, but that with richness was 0.885.

It is shown that activated sulfhydryl (SH) groups become exposed through denaturation above 76–78 °C.[58] As shown in Fig. 6.29(*b*), the response potential of ch. 1 (i.e., weakly negatively charged membrane) decreased with protein denaturation by heat treatment; the potential of ch. 7 (i.e., positively charged membrane) decreased at the same time. This result implies that negatively charged groups with the ability to bind to lipid membranes appeared at the surface of denatured proteins.

Discrimination of taste of the heat-treated samples by humans is not easy in general, and hence the usefulness of the taste sensor may further increase. In fact, five samples of milk treated at 100 °C for different processing times (0, 1, 5, 15, 30 min) were discriminated using the taste sensor. The sensory evaluations by humans showed that significant discrimination was possible only between two, the 0 min and 30 min treatments.

The detection of the degree of protein denaturation can also be applied to other fields because the measurement using the taste sensor is very convenient compared with other methods using optical equipment.

These studies were extended to examine the effect of homogenization treatment of milk on its taste. When unhomogenized milk is left for a while, it separates into two layers; the upper layer is cream and the lower part is nonfat milk because fat globules have lower specific gravity and rise. As this is unpleasant to drink, milk is homogenized to prevent the separation. Homogenization of milk is performed by using pressure to make fat globules smaller. While most of market milk is unhomogenized in some countries, homogenized milk is popular in other countries. It is often said that natural foods are more delicious than processed ones; a similar assertion is often made for unhomogenized milk. Such opinions can be examined using a scientific approach with reliable human sensory evaluations and taste-sensing devices.

To make this assessment, milk samples were homogenized at pressures of 0, 20, 70, 120, 180 and 220 kg/cm^2.[59] All the samples were prepared from the same original milk. Richness, which expresses one of the taste characteristics of milk, was evaluated by 10 well-trained panelists for the four samples with 0, 20, 120 and 220 kg/cm^2 treatments and an order of magnitude of richness was assigned to these samples. No significant difference could be detected among the four samples; for example both the 20 and 220 kg/cm² had the same richness. It implies that these samples cannot be discriminated even by well-trained panelists; the difference in taste of these milk samples must be extremely small.

Brix and electric conductivity of six milk samples were measured. Brix increased with increasing homogenization pressure from 10.9 (±0.2)% for 0 kg/cm² to 12.8 (±0.0)% for 220 kg/cm². However, no systematic change could be found for the electric conductivity, which was approximately 5.1 mS/cm, irrespective of the milk samples.

Figure 6.30 shows the fat globule size distributions in samples with 0 and 220 kg/cm² treatments measured using laser diffraction. The fat globule size became increasingly small with increasing pressure of homogenization: the median radius was 3.03, 2.49, 1.60, 1.02, 0.74 and 0.64 μm for 0, 20, 70, 120, 180 and 220 kg/cm², respectively. Only a single peak was observed for samples with pressures over 100 kg/cm², whereas two peaks appeared in samples formed at lower pressures.

Figure 6.31 shows the response patterns for six samples; Fig. 6.31(*a*) shows the samples (0, 20, 70 kg/cm²) that had two peaks in the fat globule size distribution in Fig. 6.30(*a*), while Fig. 6.31(*b*) shows the samples (120, 180, 220 kg/cm²) that had one peak in Fig. 6.30(*b*). Each sample was measured

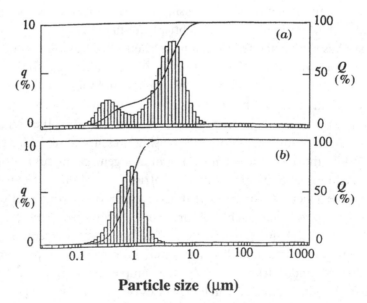

Particle size (μm)

Figure 6.30. The fat globule size distributions for 0 kg/cm² sample (*a*) and 220 kg/cm² sample (*b*).[59] The term Q is the integration of q with respect to the particle size.

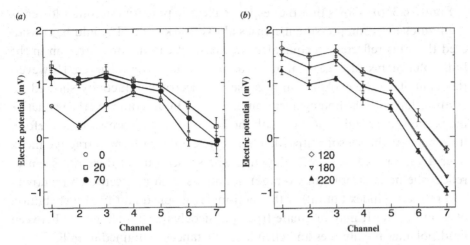

Figure 6.31. Response patterns for different homogenization-treated (kg/cm^2) samples of milk.[59]

three times, and the averages of response electric potentials were plotted together with the standard deviations. In Fig. 6.31(a), the sample 0 kg/cm^2 was distinguished from others using chs. 1 and 2. The response electric potentials in chs. 3–7 were quite similar for three samples, 0, 20 and 70 kg/cm^2, whereas the sample 0 kg/cm^2 showed the lowest response electric potentials.

In Fig. 6.31(b), three samples with 120, 180 and 220 kg/cm^2 treatments were distinguished well by all the channels in spite of very small differences of response electric potentials (0.2–0.5 mV) because of very small standard deviations of below 0.1 mV. The response electric potential decreased regularly with increasing homogenization pressure.

The PCA was applied to the data in Fig. 6.31. Milk samples were separated into two groups with homogenization pressure below and above 100 kg/cm^2 on the PC2 axis; the milk samples from 0, 20 and 70 kg/cm^2 had negative values of PC2, whereas those from 120, 180 and 220 kg/cm^2 had positive values. The contribution rates were 54.4% and 42.4% for PC1 and PC2, respectively. Since they are almost the same values and the total is 96.8%, we can discuss the response pattern safely on the two-dimensional plane of PC1 and PC2. The separation between 20 kg/cm^2 and 70 kg/cm^2 samples was not good, as also can be seen from Fig. 6.31(a). On the PC1 axis, the samples were shifted systematically to the negative direction with increasing homogenization pressure higher than 100 kg/cm^2. This implies that PC1 gives an index of change of milk quality with homogenization pressure higher than 100 kg/cm^2. PC2 reflected the situation shown in Fig. 6.30, where one or two peaks appear in the fat globule size distribution according to the homogenization pressure.

Figure 6.31(*b*) shows that the response electric potential becomes lower as the homogenization pressure increases above $100 \, kg/cm^2$. It could be considered that this reflected an equivalent decrease of cation concentration in the bulk solution because the membranes used were negatively charged. However, this cannot be the explanation because the electric conductivity showed no relationship with the homogenization pressure. On the contrary, the refractive index showed a regular increase with the homogenization pressure. Therefore, it seems that the sensor output has no correlations with the refractive index over a wide range of 0 to $220 \, kg/cm^2$. The sensor output does not seem to reflect the macroscopic (or average) quantities such as electric conductivity and refractive index but rather the semi-macroscopic quantity of distribution of globule size. It may originate from physicochemical interactions between lipid/polymer membranes and chemical substances contained in milk.

Although the fat globule size distribution had a single peak at homogenization pressures over $100 \, kg/cm^2$ (Fig. 6.30(*b*)), it had double peaks at pressures below $100 \, kg/cm^2$. When 0.01 M sodium hydroxide was used instead of ion-exchanged water as dispersion medium in the measurement of fat globule size distribution, the smaller peak of the two peaks in the lower-pressure samples disappeared. Therefore, this peak is considered to originate from casein particles, which are proteins. It can be concluded that the taste sensor clearly detects physicochemical changes in protein and lipid molecules and the molecular assemblies caused by the homogenization treatment that cannot be discriminated by humans.

Samples with different homogenization pressures were not discriminated by panelists; however, this result does not necessarily mean that taste does not vary among these samples nor does it imply that the taste sensor was not detecting taste differences as a consequence of homogenization. Some animals may have the ability to distinguish these tastes even if humans cannot. The results do suggest that the taste sensor has a superior ability over human taste in this instance. Furthermore, there is a possibility that the taste is affected by a physiochemical property such as the size of the lipid globules. Physiological studies may also be necessary for pursuing this possibility.

6.7.6 Tomatoes

In the examples discussed so far, the sensor has been applied to beverages; however, it can also be used for analysis of the taste of gelatiniform or solid foods.[60]

For quantification of the taste of tomatoes, the taste sensor was applied to commercial canned tomato juice to which four basic taste substances had been added. Data were analyzed by means of PCA. The taste of several brands of tomatoes was expressed in terms of four basic taste qualities by projecting the data obtained from these tomatoes onto the principal axes. The basic taste substances used were NaCl for saltiness, citric acid for sourness, MSG for umami taste and glucose for sweetness. No taste substance for bitterness was added because tomatoes taste only slightly bitter. Changes in response electric potentials by addition of the basic taste substances were measured as a preliminary experiment.

When eating, humans first masticate the food with their teeth and then taste it. Therefore, we used a mixer in place of teeth and crushed tomatoes before measuring them. The preconditions were established by keeping the electrode immersed in standard juice, i.e., commercial canned tomato juice without added NaCl, for two weeks. Standard juice was also used for the reference electric potential pattern, as in the case of beer.

The five kinds of tomato used were fresh-market tomatoes: Ryokken, Kiss, Fukken, TVR-2 and a kind of tomato for processing (hereafter named PT). Ryokken and Kiss were cultivated to be sweeter.

Figure 6.32 shows the response patterns for several samples of one brand, TVR-2. Each sample was measured more than five times. Typical standard deviations were 0.17, 0.49, 0.37, 0.34, 0.40, 0.27, 0.68 and 0.14 mV for chs. 1–8. The patterns were measured relative to the standard juice. The magnitude of response was very different for each sample. However, the response patterns resembled each other in their shapes. The same situation was also observed for the other brands.

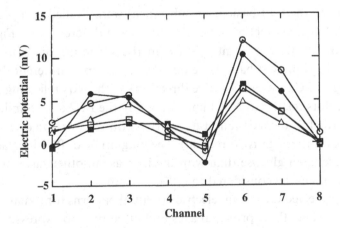

Figure 6.32. Output patterns for several samples of TVR-2.[60] Different symbols denote different samples.

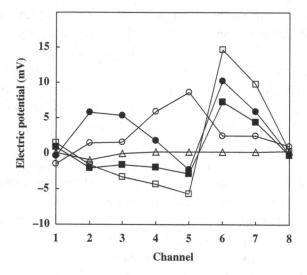

Figure 6.33. Output patterns for five brands of tomatoes:[60] TVR-2 (●); Fukken (○); Ryokken (■); Kiss (□); and PT (△).

Figure 6.33 shows examples of the response pattern for one sample each of five brands of tomatoes. Different brands of tomatoes were distinguished by the shapes of the output electric potential patterns. Therefore, tomatoes of the same brand can be considered to have a taste with similar proportions; the difference in taste among tomatoes of the same brand may result mainly from the difference in magnitude of taste, because the output electric potential changed linearly with the concentrations of taste substances in a narrow range, as also seen in Fig. 6.5.

Responses to the four basic taste substances were studied as a preliminary experiment. Each channel responded to them in a different way. Figure 6.34 shows the obtained data points plotted in the scattering diagram of PCA. Dots indicating different basic taste qualities lined up in different directions. We can regard PC2 (in the negative direction) as the axis reflecting saltiness, because the dots related to NaCl lined up along the PC2 axis. Similarly, PC1 can be regarded as the axis reflecting sourness (in the positive direction) and umami taste (in the negative direction). The magnitude of PC3 increased more with the increase in glucose than with the increase in other taste substances; hence this axis can be considered to reflect sweetness.

Then we projected the output electric potential patterns for tomatoes in Fig. 6.33 onto the same three principal axes and attempted to express taste characteristics of each tomato in terms of the four basic taste qualities. Figure 6.35 shows the results of projection using the transformation matrix obtained in

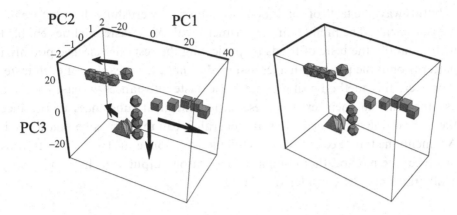

Figure 6.34. Scattering diagram of PCA.[60] Citric acid, NaCl, glucose and MSG are shown by the symbols, cube, tetrahedron, dodecahedron and icosahedron, respectively. The arrow indicates the increase in concentration of each chemical. The units for PC1, PC2 and PC3 are mV.

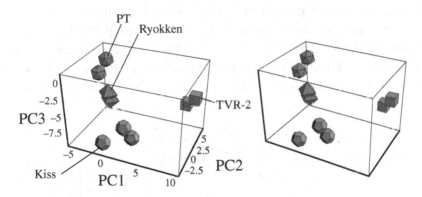

Figure 6.35. Projection of the response patterns for tomatoes to the three principal component axes.[60] Different points with the same symbol indicate different samples of the same brand of tomato. The units for PC1, PC2 and PC3 are mV.

Fig. 6.34. We can predict the taste of each tomato as follows. The PC1 scores for TVR-2 tend to be so positively high that TVR-2 can be considered to taste sour. This result agrees with the human sensation, because TVR-2 is a relatively old brand of tomato and is generally acknowledged as a sour tomato. For sweet tomatoes such as Ryokken or Kiss, the PC3 scores tend to be negatively high; hence these tomatoes are presumed to be sweet, as expected. The PC1 scores for most tomatoes are positively higher than PT; hence PT is presumed to have a relatively strong umami taste. Tomatoes tasting strongly of umami are generally used for processing; this prediction also agrees with human sensation.

In this way, the taste of solid foods was studied by crushing them before the measurements. The preliminary experiment with basic taste qualities enabled us to quantify the taste of foods by projecting the response electric potential patterns onto the plane with axes assigned to have the meaning of each taste. However, this method using some basic taste substances is based on the assumption that there exist representative chemical substances to produce the taste of the concerned food. In general, it cannot always be guaranteed. As mentioned in Section 6.7.3, caffeine-free coffee tastes bitter. In this instance, the method that compares the sensor output with human sensory evaluations may be considered as better.

6.7.7 Rice

The eating quality of rice is evaluated by attributes such as appearance, flavor, stickiness, hardness and taste. Attributes except for taste have been measured using some apparatus such as colorimeter, gas chromatograph, viscosity meter and near-infra-red analyzer. Taste is judged by humans. To assess the taste of rice using the taste sensor,[61] samples were prepared by mixing rice with water with weight ratio 1 : 10 and then boiling it. Six Japanese domestic rices (Koshihikari from different producing districts) were used and each type of rice was distinguished by the output patterns of the taste sensor. Figure 6.36

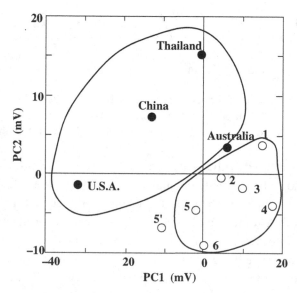

Figure 6.36. PCA applied to the response patterns for rice.[61] The numbers 1–6 refer to Japanese domestic rice. The domestic rice 5 was assayed fresh and after storage for one month in sunshine (5′).

shows the result of PCA applied to these data. The rice could be clearly classified into two groups, Japanese domestic and non-Japanese rice. Australian rice was closest in character to the domestic rices. In sensory evaluations of cooked rices, the taste (Y) evaluated by humans was found to have a relationship with PC1 and PC2 from the sensor output ($\hat{r} = 0.80$) by

$$Y = 0.046 \times \text{PC1} - 0.050 \times \text{PC2} - 1.06. \tag{6.21}$$

Furthermore, Fig. 6.36 shows that fresh rice (denoted by 5) is distinguished from rice (denoted by $5'$) that was stored for one month in sunshine. This implies that the change of quality of rice with time can be detected using the taste sensor.

This study shows that the multichannel taste sensor can be used to distinguish the variations in taste in rice.

6.8 Quality control of foods

6.8.1 Sake

The quality of sake is evaluated by factors such as taste, smell and color. The evaluation of taste is the most difficult among these factors. To date, the evaluation of the taste of sake has been carried out through chemical analyses in which the relationship between chemical substances and taste has been investigated. However, sensory tests by humans are considered more important, although they provide no quantitative evaluation.

Samples for measurements using the taste sensor are as follows.[62]

Junmaishu This is made from a combination of highly refined polished rice (polished up to 70% of its original size), koji (fermented rice) and fresh stream water. Sugar and alcohol are not added.

Daiginjoshu This is made from highly refined polished rice (polished up to 50%) that is slowly fermented at low temperatures. As a result this sake has a rich and fruity flavor.

Honjozoshu Polished rice (polished up to 70%), koji, water and alcohol are blended together using traditional techniques to give this sake a unique color and a rich, smooth aroma. It most resembles ordinary sake among the four varieties.

Alps ginjoshu This is an original sake made in Nagano Prefecture. It resembles Daiginjoshu except for the degree of rice polishing (polished up to 60%) and the alcohol density (low-alcohol sake: diluted with water).

Figure 6.37 shows sensor responses for four sakes from one brewery, measured by adopting an ordinary sake from the same brewery as the standard.[62]

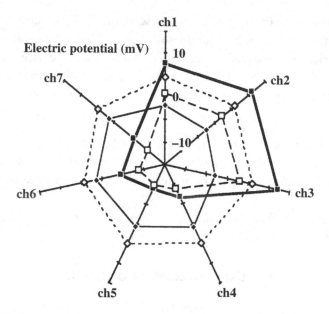

Figure 6.37. Response electric potential patterns for various types of sake from one brewery.[62] Daiginjoshu (■), Junamishu (□), Honjozoshu (◆), and Alps ginjuoshu (◇).

The same lipid/polymer membranes were used in chs. 1, 2 and 3 (charged negatively), chs. 4 and 5 (charged positively), and chs. 6 and 7 (intermediate positive electric charge).

As found in Fig. 6.37, the responses of these three membrane types to the four sakes were different. Distinction of the sakes was very easy because of the low standard deviations, e.g., ch. 1 (0.16 mV), ch. 4 (0.31 mV) and ch. 6 (0.25 mV) in the analysis of Honjozoshu, repeated 40 times.

When highly alcoholic drinks such as sake are measured, the durability and stability of the lipid/polymer membrane are important. Figure 6.38 shows the changes in response to Honjozoshu with repeated use of the same electrode. Every membrane was stable for more than one month. The changes in response of three membranes to Honjozoshu were within 1 mV. Actually, the same electrode has been safely used for more than two years ; the similar situation also holds in multichannel electrodes for other foodstuffs.

Figure 6.39 shows the changes in pH, the titratable acidity and the response electric potential of ch. 1 in the fermentation process of sake.[63] The pH decreased and then increased: consequently, this cannot be used as a measure of fermentation. However, the titratable acidity can be used as the measure because it increased regularly. The sensor response increased in a similar way.

Multiple regression analysis was performed on the titratable acidity using the two sensor outputs in the fermentation process of sake. The correlation

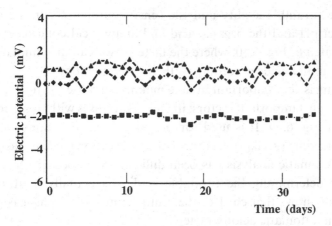

Figure 6.38. Change in responses of chs. 1, 4 and 7 to sake with repeated use of the same electrode.[62]

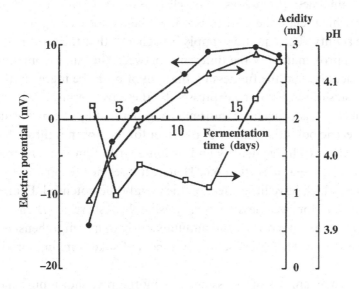

Figure 6.39. Changes in pH (□), the titratable acidity (△) and the response electric potential of ch. 1 (●) in the fermentation process of sake.[63]

was very high: 0.99. The degree of titratable acidity is influenced by the pH of samples and by the concentrations of organic acids and amino acids. Increasing organic and amino acid concentrations induce greater buffer action, which results in an increase of titratable acidity. The correlation between the response of ch. 1 and a function of pH, organic acid concentration and amino acid concentration was as high as 0.95. The high correlation between the titratable acidity and the sensor outputs occurs because both values reflect pH and the organic acid and amino acid concentrations. This is demonstrated in Fig. 6.10, where the taste sensor can discriminate between the taste of amino acids according to their taste qualities.

This finding is very important in sake making, because the titratable acidity is one of the most important factors; in fact, it increases with the fermentation, as shown in Fig. 6.39. It is used for the adjustment of taste sweetness and thickness, blending of sake and control of fermentation of sake mash moromi. However, automatic analysis has been difficult so far because the analytical method for determining the titratable acidity is actually a titration. The present result shows that ch. 1 of the multichannel taste sensor is promising for use as an automatic acidity meter.

The taste of sake is quite delicate and complicated. However, Sato *et al.*[64] showed that the taste could be adequately represented by three major factors, sweetness, thickness (or fullness) and cleanness. Moreover, they[65] proposed equations which relate the taste to two ingredients: the sugar content and the titratable acidity. The equations imply, in general, that the sweetness of sake could be approximated by the difference between the sugar content and the titratable acidity, and the fullness by the sum of both the ingredients. Hence, the two tastes of sake may be expressed by these two terms.

The sensor transforms taste information generated by chemical substances into the electric potential change of the receptor membranes through changes of surface electric charge density and/or ion distribution near the membrane surface, as discussed in Section 6.6. Hence, it does not respond well to non-electrolytes, which have little effect on the membrane potential. Evaluation of the taste of sake for sweetness is not easy using the taste sensor; hence, we used here an enzymatic glucose sensor simultaneously in a comprehensive evalua-tion of the sake taste,[66] because the sweetness of sake is mainly produced by glucose.

In the early examples of assessing beer by the preconditioning method, a particular commercial brand was used as a standard solution. This is not necessarily a good method because the same brand with the same lot (e.g., production date, factory) is not always available. Consequently, a standard solution of sake was made for the sake assessments. This reduced experimen-tal difficulties because the solution was always available and could be replaced at any time. The standard solution was prepared after some trials by taking account of a synthetic sake.[64] Its ingredients are as follows (amounts per liter): glucose 20.0 g, starch syrup 5.55 g, succinic acid 0.50 g, sodium succinate 0.22 g, MSG 0.22 g, glycine 0.11 g, L-alanine 0.11 g, NaCl 0.15 g, KH_2PO_4 0.06 g, $CaHPO_4$ 0.06 g, tyrosol 2.00 g and ethanol 150 ml.

Seventeen brands of Japanese sake were prepared: Hana no Tsuyu Ryosen (1), Hana no Tsuyu Ginjo (2), Kiku Tama no I (3), Ume Nishiki Akiokoshi (4), Suishin Namaginjo (5), Shiratsuyu Honjozo (6), Gekkeikan (7), Ginban Namaginjo (8), Chiyo no Haru Junmai Ginjo (9), Tsuruhime Ginjo (10), Hakutsuru Kizake (11), Chiyo no Haru Namachozoshu (12), Hakutsuru Josen (13), Wakatake (14), Reiho (15), Yoshinoyama (16) and Hakutsuru Namachozoshu (17).

For quantitative scales for acidity and sugar content, we prepared comparative solutions that contain different concentrations of succinic acid and glucose, respectively, in the standard solution. This method is similar to that used in the tomato studies in Section 6.7.6. The concentrations (amounts per liter) of succinic acid were chosen as 0.125, 0.25, 0.5, 1.0 and 2.0 g (series 1), and those of glucose 5, 10, 15, 20, 25, 30 and 35 g (series 2). The comparative solutions and commercial sake were measured with the taste sensor using the standard solution as reference. PCA was applied to data obtained for the solutions of series 1 using the taste sensor. Factor loadings (a_{1i} for $i = 1 \dots 8$) to PC1 were obtained by this process. Let us assign the transformation coefficient of the output of the glucose sensor for series 2 to a_{2g}, then the transformation matrix M from the data by the taste sensor and glucose sensor defined by

$$\begin{pmatrix} T_a \\ T_s \end{pmatrix} = M \begin{pmatrix} v_1 \\ \vdots \\ v_8 \\ v_g \end{pmatrix}, \tag{6.22}$$

leads to

$$M = \begin{pmatrix} a_{11}, a_{12}, \dots, a_{18}, & 0 \\ 0, & a_{2g} \end{pmatrix}. \tag{6.23}$$

Here, T_a and T_s are the abscissa and ordinate of the transformed two-dimensional plane, respectively, $v_1 \dots v_8$ designating the output from chs. 1–8 of the taste sensor, and v_g the output of the glucose sensor, where the average was subtracted from the output so that the response potential for the standard solution might be located at the origin. In the present case, M has the explicit form:

$$M = \begin{pmatrix} 0.00 & -0.05 & 0.49 & 0.22 & 0.08 & 0.36 & 0.52 & 0.54 & 0 \\ 0 & 0 & 0 & 0 & 0 & 0 & 0 & 0 & 1 \end{pmatrix}. \tag{6.24}$$

Using eq. (6.22) with eq. (6.24), response potentials for several brands of commercial sake were transformed to the two-dimensional plane (T_a, T_s). Results are summarized in Fig. 6.40.

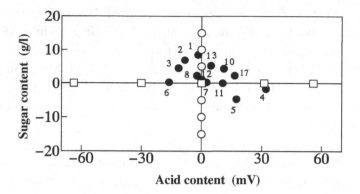

Figure 6.40. Taste map of Japanese sake.[66] ● Commercial sake; ○, artificial solutions with different glucose concentrations; □, solutions with different succinic acid concentrations. Attached numbers indicate different brands of sake.

When the standard solution was used as reference, all the brands of commercial sake are found to be distributed around the origin of T_a and T_s axes in Fig. 6.40. The abscissa and ordinate axes imply the acid content and the sugar content, respectively. Therefore, both the contents of acid and sugar in sake can be easily found with this map. The correlation coefficient was 0.999 between the abscissa T_a and the logarithm of succinic acid concentration in Fig. 6.40. It implies that the abscissa (mV) can be easily transformed into the succinic acid concentration (g/l). Since other acid components are included in commercial sake, the abscissa of the taste map in Fig. 6.40 can be interpreted as the equivalent concentration in terms of succinic acid.

According to Sato *et al.*,[65] the sweetness is approximated by the difference between sugar and titratable acidity, and the thickness by the sum of both ingredients:

$$[\text{sweetness}] = 0.86 \times [\text{sugar content}] - 1.16 \times [\text{titratable acidity}] + \text{const.}$$
$$[\text{thickness}] = 0.42 \times [\text{sugar content}] + 1.88 \times [\text{titratable acidity}] + \text{const.}$$

(6.25)

Therefore, sake in the upper right section of the map should have a sweet–thick taste, sake in the lower right section shows dry–thick taste.

6.8.2 Soybean paste

Miso (soybean paste) is a traditional, fermentative food in Japan. One of the original materials, so-called koji (cultivated rice or barley), is made by adding seeds of koji fungus to steamed rice or barley. As the second step, koji, boiled soybean and salt are mixed. The immature mixture is put into a big tank for fermentation. When the fermentation finishes, it produces a delicious, pre-servative foodstuff miso.

The degree of fermentation of miso, that is to what extent the mixture of koji, soybean and salt is ripe, has been mainly judged by human experience. This is influenced by the daily physical conditions of the individual and by the environment. Therefore an objective measure that is not influenced by any environment has been desired for a long time.

However, some chemical analyses, for example for titratable acidity, are used to monitor miso fermentation. There is a good correlation between the titratable acidity and the degree of fermentation. Amino acids also increase as fermentation progresses. However, these analyses are laborious and time con-suming; therefore, they are unsuitable for real-time or on-line measurement in miso fermentation. A simpler and more useful way to judge the miso fermen-tation is necessary for assurance of reliable quality and automatized produc-tion of miso.

Figure 6.41 shows the time courses of the responses of three channels, which are made of DOP, TOMA and OAm, of the taste sensor to miso with fer-mentation and storage days.[67] This result indicates that the DOP and OAm membranes responded linearly to samples in the fermentation process (0–40 days), while the changes of response to ripe samples that were shipped (50–320 days) were smaller than those in samples in the fermentation process. The TOMA membrane showed a small, gradual increase with days. This

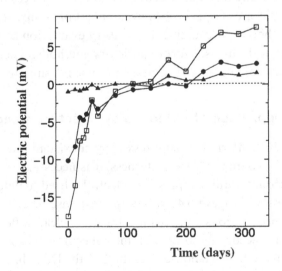

Figure 6.41. Changes in responses of DOP (●), TOMA (▲) and OAm (□) membranes for miso with fermentation and storage days.[67]

result suggests that the DOP and OAm membranes can be used as monitoring of the fermentation of miso.

As can be found in Fig. 6.41, the response potentials became positive beyond about 150 days (about 5 months). This fact is interesting because it is normal for miso products on sale beyond 5 months to be returned to the factory. Thus, the taste sensor can provide a criterion of reliable quality of miso products.

It was also found that the response potentials of the DOP and OAm membranes reflected the amount of amino acids such as L-glutamic acid, L-leucine (including L-isoleucine), L-aspartic acid and L-arginine. Furthermore, the correlation coefficients between the sensor outputs of the DOP and OAm membranes and titratable acidity were 0.87 and 0.88, respectively. The titratable acidities are affected by not only the pH of the samples but also the concentrations of organic acids and amino acids. Therefore, the increasing titratable acidities in the fermentation process are compatible with the increasing amino acid concentrations. These physicochemical quantities could form the basis for a potential method to monitor the process; so far, however, this has not been achieved because of the intricacy of the analysis methods. Therefore, analysis using the taste sensor is very feasible because of its convenience for use as an on-line measurement of miso fermentation.

6.9 Suppression of bitterness

It is important especially for pharmaceutical and food sciences to express the extent of bitterness, for example in the development of syrups. To date, however, the main method of measurement is sensory evaluation by humans and conventional chemical analyses form subsidiary methods. Therefore, taste-sensing devices that could detect bitterness have been desired for a long time.

6.9.1 Suppression of bitterness by sweet substances

It is well known that bitterness is suppressed by coexistent sweet substances such as sucrose.[7,68] To quantify the bitterness, it is necessary to measure the taste by taking account of this suppression effect. We tried to detect this effect using the taste sensor.[69] Figure 6.42 shows the response patterns for quinine with different concentrations and demonstrates a clear difference of the response potentials, according to the quinine concentrations. As the quinine concentration increased, the response potentials of the DA (ch. 1), OA (ch. 2), DOP (ch. 3), DOP : TOMA 5 : 5 (ch. 4) and 3 : 7 (ch. 5) membranes increased, whereas those of the TOMA (ch. 6) and OAm (ch. 7) membranes decreased.

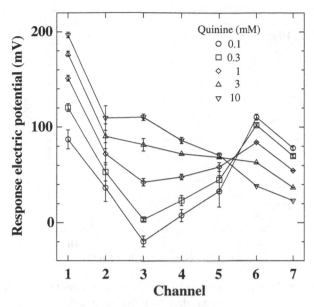

Figure 6.42. Response patterns for quinine.[69]

The increase in chs. 1–5 is because quinine hydrochloride becomes positively charged with ionization and is adsorbed strongly to the membranes, with its hydrophobicity changing the membrane electric charge from negative to positive, as explained quantitatively in Fig. 6.19. The decrease in chs. 6 and 7 is brought about by Cl⁻. Since most bitter substances are hydrophobic, the behavior observed in chs. 1–5 can be considered to reflect well the characteristics of bitterness, as also discussed for amino acids in Fig. 6.11.

The measurement was also made on samples of 1 mM quinine with sucrose of different concentrations. The response pattern shifted downward as a whole with increasing sucrose concentration. The decrease in response potentials of chs. 1–5 was as large as 10 mV or more and was the opposite to the above change in response potentials for increasing quinine concentration. Therefore, it may be possible to regard this decrease as the equivalent of a decrease in quinine concentration (or decrease in bitterness) with increasing sucrose concentration.

Application of PCA to these data to quantify bitterness is shown in Fig. 6.43, the relationship between PC1 and the quinine concentration. The contribution rates of the original data to PC1, PC2 and PC3 were 95.4%, 3.4% and 0.7%, respectively. This means that PC1 characterizes the patterns in Fig. 6.42 and the data, can be analyzed using only PC1. The PC1 is expressed by the formula

$$\mathrm{PC1} = a_1(v_1 - v_1') + a_2(v_2 - v_2') + \cdots + a_7(v_7 - v_7') \tag{6.26}$$

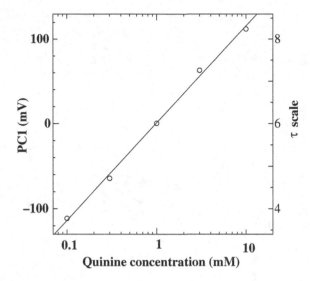

Figure 6.43. Relationship between PC1 data and the quinine concentration.[69]
The τ scale is derived using eq. (6.29).

where a_i is the factor loading, v_i the response potential of ch. *i*, and v_i' is the response potential of ch. *i* averaged over the measured samples. In the present case, a_i and v_i' were determined as

$$a_i = (0.477, \ 0.324, \ 0.590, \ 0.358, \ 0.172, \ -0.317, \ -0.249),$$
$$v_i' = (146.4, \ 72.39, \ 43.81, 47.57, 55.12, 79.70, \ 52.54). \tag{6.27}$$

In Fig. 6.43, a straight line is drawn using the method of least squares. It can be seen that PC1 increases in proportion to the quinine concentration in logarithmic scale. It is noticeable that this relation agrees with the well-known Weber–Fechner's law of human sensory evaluation.[7] This law states that a sensation is proportional to the logarithm of stimulus intensity.

The result of Fig. 6.43 means that PC1 can be considered to express the strength of bitterness. Therefore, we can transform the PC1 value obtained from the sensor output to the τ scale as the measure of bitterness to express the reported relationship between the τ scale and quinine concentration:

$$\tau = 2.35 \log(c/0.00011), \tag{6.28}$$

where *c* is the quinine concentration (g/100 cm^3). The τ scale represents the subjective strength of a taste solution. The least squares line in Fig. 6.43 is given by

$$PC1 = 114.75 \log C + 1.05 \tag{6.29}$$

with C denoting the molar concentration; hence, we get a relationship between τ and PC1:

$$\tau = 0.0205\text{PC1} + 5.98. \tag{6.30}$$

If the PC1 value is calculated we can estimate the strength of bitterness expressed by the τ scale or the equivalent quinine concentration c given by eq. (6.28). The strength of bitterness was then obtained from the response potentials for samples that contained sucrose, using eqs. (6.26), (6.27) and (6.30): The result is shown as a function of the sucrose concentration in Fig. 6.44. There is a decrease of the strength of bitterness with increasing sucrose concentration despite a constant quinine concentration of 1 mM. This result suggests that a satisfactory detection of the suppression of bitterness induced by sucrose is achieved with a taste sensor.

This procedure has also been used to assess the effect of sucrose in decreasing the bitterness of a drug.[69]

6.9.2 Suppression of bitterness by phospholipids

A lipoprotein (PA-LG) made of phosphatidic acid (PA) and β-lactoglobulin (LG) completely suppressed bitterness without affecting other taste qualities.[70] A similar but a little weaker suppressive effect was found using only PA, which may have a structure similar to that of a liposome in aqueous

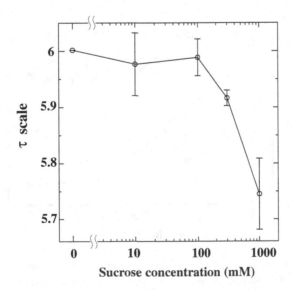

Figure 6.44. Decrease of bitterness of quinine as a consequence of increasing coexistent sucrose expressed by the τ scale.[69]

solution; addition of 1% PA to the solutions containing bitter substances such as quinine and propranolol decreased the bitterness to the extent that humans sensed no or a very weak bitter taste.[71]

This suppression effect was studied[72] using the taste sensor and a commercial bitter-masking substance (BMI-60) composed of phospholipids; its ingredients are 15–20% PA, 5% phosphatidylcholine, 40% phosphatidylinositol and 10–15% phosphatidylethanolamine. The measurement and data processing methods were the same as those described above.

Figure 6.45 shows the change in equivalent quinine concentration with increasing concentration of the above phospholipids. Two quinine concentrations of 0.1 mM and 1 mM were studied, and two methods of application of the phospholipids to the quinine solution were tried. One method is the measurement of the mixed solution of quinine and the phospholipids, and another is the measurement of quinine solution using an electrode that had been pretreated by immersing it in a solution containing the phospholipids for 10 s. The former method permits the possibility that the phospholipids directly interact with quinine, while the latter contains the modification of

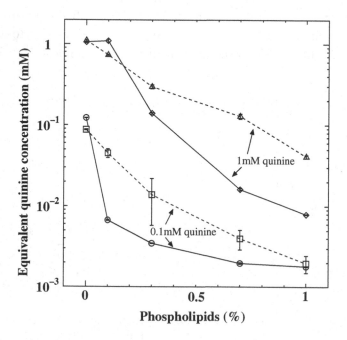

Figure 6.45. Change of the equivalent quinine concentration as a result of addition of phospholipids.[72] Assessment using a solution of mixed quinine and phospholipids is denoted with a solid line, while that for quinine solution using the phosholipid-modified electrode is denoted by the dashed line.

the lipid/polymer membranes owing to the presumable adsorption of the phospholipids.

Using a mixed solution containing 0.1 mM quinine, addition of 0.1% phospholipids decreases the equivalent quinine concentration to about 8 μM which elicits no bitter taste. For the solution of 1 mM quinine, 0.7% and 1% addition resulted in a weak or absent bitterness. In the second method where the quinine solution is measured using the lipid/polymer membranes modified by adsorption of the phospholipids, the bitterness of 0.1 mM quinine was effectively weakened to the zero level beyond 0.3% phospholipids. The bitterness of 1 mM quinine was also mostly masked.

The present method using the taste sensor can be expected to provide a new automated method to measure the strength of the bitterness of a drug substance to replace human sensory evaluation. In addition, study of the taste interaction using the taste sensor will contribute to clarification of reception mechanisms in the gustatory system.

6.10 Taste-sensing field effect transistors

As mentioned above, taste can be quantified using a multichannel taste sensor with lipid/polymer membranes. Its sensitivity and stability are superior to taste assessment by humans.

Depending on the purpose and objective to be measured, it may be necessary to miniaturize the taste sensor. In fact, the taste sensor can be applied to real-time measurements of chemical substances in organisms as well as in foodstuffs. The studies described below are the first step in development of an integrated taste-sensing FET (TSFET).[73]

Figure 6.46 shows the detecting part made of MOSFET (metal-oxide semiconductor FET), on the gate of which the lipid/polymer membrane is pasted. The size of the FET is 1 mm width, 2 mm length and 0.5 mm thickness, and the gate part has an open 10 μm $\times$ 340 μm area. The membrane was made in the same way as above. The membrane of 1 mm $\times$ 2 mm size was pasted on the gate using 1% PVC/THF solution, which inhibited easy detachment of the membrane. Eight MOSFET electrodes with different lipid/polymer membranes were separately prepared, and the gate-source voltage was measured by keeping the source-drain current constant. The output was taken into a computer using a handmade scanner for exchanging sequentially the eight outputs.

Four foodstuffs, coffee, canned coffee, beer and a soft drink, were measured. These foodstuffs were discriminated well with very small standard

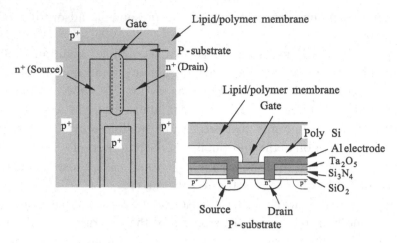

Figure 6.46. Detecting part of MOSFET, on the gate of which the lipid/polymer membrane is pasted.[73]

deviations, e.g., 0.82, 0.47, 0.47 and 0.83 mV in DOP, OAm, TOMA and DOP:TOMA 5:5 membranes, respectively, for the soft drink.

The FET taste sensor was found to show the same characteristics as the conventional taste sensor. It is because the same lipid/polymer membranes are used: the output (V_{gs}) is the change in membrane potential brought about by interactions between the membrane and chemical substances. However, the stability was not as good as that of the conventional taste sensor; the lipid/polymer membranes detached from the MOSFET after use of over 100 h, although the 1% PVC/THF treatment was very effective in prolonging the lifetime, as already mentioned. Without the PVC/THF treatment, a stable response potential was not obtained sometimes from the first attempt.

A dihexadecyl phosphate (DHP) LB membrane was also used. The PVC membrane was first formed on the FET surface using a casting method to make hydrophobic chains of DHP stably attached to the gate part. As a result, responses to chemical substances such as quinine and HCl differed from the responses of the lipid/polymer membranes above.

In this study, an effective integrated taste sensor was not made but its production should be possible based on the results achieved. This will then allow fabrication of an integrated taste sensing system.

REFERENCES

1. Hayashi, K., Yamanaka, M., Toko, K. and Yamafuji, K. (1990). *Sens. Actuators*, B2, 205.
2. Toko, K., Hayashi, K., Yamanaka, M. and Yamafuji, K. (1990). *Tech. Digest 9th Sens. Symp.*, p. 193.

3. Kikkawa, Y., Toko, K., Matsuno, T. and Yamafuji, K. (1993). *Jpn. J. Appl. Phys.*, 32, 5731.
4. Mikhelson, K. N. (1994). *Sens. Actuators*, B18-19, 31.
5. Watanabe, M., Toko, K., Sato, K., Kina, K., Takahashi, Y. and Iiyama, S. (1998). *Sens. Mater.*, 10, 103.
6. Toko, K., Nakagawa, Y., Obata, M. and Yahiro, T. (1997). *Sens. Mater.*, 9, 297.
7. Pfaffmann, C. (1959). In *Handbook of Physiology*, Sect. 1 *Neurophysiology*, Vol. 1, ed. Field, J. American Physiological Society, Washington, DC, p. 507.
8. Yamaguchi, S. and Takahashi, C. (1984). *Agric. Biol. Chem.*, 48, 1077.
9. Ninomiya, Y. and Funakoshi, M. (1987). In *Umami: A Basic Taste*, eds. Kawamura, Y. & Kare, M. R. Marcel Dekker, New York, p. 365.
10. Ikezaki, H., Toko, K., Hayashi, K., Toukubo, R., Yamanaka, M., Sato, K. and Yamafuji, K. (1991). *Tech. Digest 10th Sens. Symp.*, p. 173.
11. Watanabe, T., Kawada, T. and Iwai, K. (1987). *Agric. Biol. Chem.*, 51, 75.
12. Hagerman, A. E. and Butler, G. (1981). *J. Biol. Chem.*, 256, 4494.
13. McManus, J. P., Davis, K. G., Lilley, T. H. and Haslam, E. (1981). *J. Chem. Soc., Chem. Commun.*, 7, 309.
14. Bate-Smith, E. C. (1954). *Food Proc. Indust.*, 23, 124.
15. Lea, A. G. H. and Arnold, G. M. (1978). *J. Sci. Food Agric.*, 29, 478.
16. Lyman, B. J. and Green, B. G. (1990). *Chem. Senses*, 15, 151.
17. Kawamura, Y., Funakoshi, M., Kasahara, Y. and Yamamoto, T. (1969). *Jpn. J. Physiol.*, 19, 851.
18. Schiffman, S. S., Suggs, M. S., Sostman, A. and Simon, S. A. (1991). *Physiol. Behav.*, 51, 55.
19. Iiyama, S., Toko, K., Matsuno, T. and Yamafuji, K. (1994). *Chem. Senses*, 19, 87.
20. Szolcsanyi, J. (1977). *J. Physiol.*, 77, 251.
21. Dunér, M.-E., Fredholm, B. B., Larsson, O., Lundberg, J. M. and Saria, A. (1986). *J. Physiol.*, 373, 87.
22. Joslyn, M. A. and Goldstein, J.L. (1964). *Adv. Food Res.*, 13, 179.
23. Toko, K. and Fukusaka, T. (1997). *Sens. Mater.*, 9, 171.
24. Nagamori, T. and Toko, K. (1999). *Dig. Tech. Papers Transducers '99*, p. 62.
25. Toko, K. and Nagamori, T. (1999). *Trans. IEE Jpn*, 119-E, 528.
26. Ninomiya, T., Ikeda, S., Yamaguchi, S. and Yoshikawa, T. (1966). *Proc. 7th Congress Sensory Tests*, p. 109 [in Japanese].
27. Birch, G. G. (1987). In *Umami: A Basic Taste*, eds. Kawamura, Y. & Kare, M. R. Marcel Dekker, New York, p. 173.
28. Kirimura, J., Shimizu, A., Kimizuka, A., Ninomiya, T. and Katsuya, N. (1969). *J. Agr. Food Chem.*, 17, 689.
29. Ney, K. H. (1971). *Z. Lebensm.-Unters. Forsch.*, 147, 64; Tanford, C. (1962). *J. Am. Chem. Soc.*, 84, 4240.
30. Toko, K., Matsuno, T., Yamafuji, K., Hayashi, K., Ikezaki, H., Sato, K., Toukubo, R. and Kawarai, S. (1994). *Biosens. Bioelectron.*, 9, 359.
31. Pangborn, R.M. (1960). *Food Res.*, 25, 245.
32. Indow, T. (1969). *Percept. Psychophys.*, 5, 347.
33. Murata, T., Hayashi, K., Toko, K., Ikezaki, H., Sato, K., Toukubo, R. and Yamafuji, K. (1992). *Sens. Mater.*, 4, 81.
34. Oohira, K. and Toko, K. (1996). *Biophys. Chem.*, 61, 29.
35. Oohira, K., Toko, K., Akiyama, H., Yoshihara, H. and Yamafuji, K. (1995). *J. Phys. Soc. Jpn*, 64, 3354.
36. Kamo, N. and Kobatake, Y. (1974). *J. Colloid Interface Sci.*, 46, 85.
37. Träuble, H., Teubner, M., Woolley, P. and Eibl, H. (1976). *Biophys. Chem.*, 4, 319.
38. Ohshima, H. and Mitsui, T. (1978). *J. Colloid Interface Sci.*, 63, 525.
39. Payens, Th. A. J. (1955). *Philips Res. Rep.*, 10, 425.
40. Goldman, D. E. (1943). *J. Gen. Physiol.*, 27, 37.

41. Kobatake, Y., Kurihara, K. and Ueda, T. (1975). *Physical and Chemical Basis of Life II*. Iwanami Syoten, Tokyo, p. 337 [in Japanese].
42. Kobatake, Y. (1975). *Adv. Chem. Phys.*, 29, 319.
43. Kamo, N., Yoshioka, T., Yoshida, M. and Sugita, T. (1973). *J. Membr. Biol.*, 12, 193.
44. Nomura, K. and Toko, K. (1992). *Sens. Mater.*, 4, 89.
45. Toko, K., Murata, T., Matsuno, T., Kikkawa, Y. and Yamafuji, K. (1992). *Sens. Mater.*, 4, 145.
46. Ikezaki, H., Hayashi, K., Yamanaka, M., Tatsukawa, R., Toko, K. and Yamafuji, K. (1991). *Trans. IEICE Jpn*, J74-C-II, 434 [in Japanese].
47. Kirin Brewery Co., Ltd (ed.) (1992). *Biru no Umasa wo Saguru* (*Study of Deliciousness of Beer*). Shokabo, Tokyo, p. 81 [in Japanese].
48. Ezaki, S., Yuki, T., Toko, K., Tsuda, Y. and Nakatani, K. (1997). *Trans. IEE Jpn*, 117-E, 449 [in Japanese].
49. Wada, M. (1994). *Quark*, 13, 40 [in Japanese].
50. Toko, K. (1998). *Sensors Update*, Vol. 3, eds. Baltes, H., Göpel, W. & Hesse, J. Wiley-VCH, Weinheim, p. 131.
51. Iiyama, S., Yahiro, M. and Toko, K. (1995). *Sens. Mater.*, 7, 191.
52. Taniguchi, A., Naito, Y., Maeda, N., Ikezaki, H. and Toko, K. (1998). *Trans. IEE Jpn*, 118-E, 1634 [in Japanese].
53. Di Natale, C., Davide, F., Brunink, J. A. J., D'Amico, A., Vlasov, Y. G., Legin, A. V. and Rudnitskaya, A. M. (1996). *Sens. Actuators*, B34, 539.
54. Fukunaga, T., Toko, K., Mori, H., Nakabayashi, T. and Kanda, M. (1996). *Sens. Mater.*, 8, 47.
55. International Coffee Organization (1991). *Consumer-Oriented Vocabulary for Brewed Coffee*, Report No. 3, p. 4.
56. Toko, K., Iyota, T., Mizota, Y., Matsuno, T., Yoshioka, T., Doi, T., Iiyama, S., Kato, T., Yamafuji, K. and Watanabe, R. (1995). *Jpn. J. Appl. Phys.*, 34, 6287.
57. American Dry Milk Institute (1971). *Standards for Grades of Dry Milk Including Methods of Analysis*. American Dry Milk Institute, Illinois, p. 4.
58. Tsugo, T. and Yamauchi, K. (1975). *Gyunyu no Kagaku* (*Chemistry of Milk*), Chikyu-sha, Tokyo, Ch. 5 [in Japanese].
59. Yamada, H., Mizota, I., Toko, K. and Doi, T. (1997). *Mater. Sci. Eng.*, C5, 41.
60. Kikkawa, Y., Toko, K. and Yamafuji, K. (1993). *Sens. Mater.*, 5, 83.
61. Yahiro, M., Toko, K. and Iiyama, S. (1997). *Trans. IEE Jpn*, 117-E, 187 [in Japanese].
62. Arikawa, Y., Toko, K., Ikezaki, H., Shinha, Y., Ito, T., Oguri, I. and Baba, S. (1995). *Sens. Mater.*, 7, 261.
63. Arikawa, Y., Toko, K., Ikezaki, H., Shinha, Y., Ito, T., Oguri, I. and Baba, S. (1996). *J. Ferment. Bioeng.*, 82, 371.
64. Sato, S., Tadenuma, M., Takahashi, K., Azuma, Y., Tanzawa, K. and Edamura, H. (1974). *J. Soc. Brew. Jpn*, 69, 771.
65. Sato, S., Kawashima, H. and Murayama, Y. (1974). *J. Soc. Brew. Jpn*, 69, 774.
66. Iiyama, S., Suzuki, Y., Ezaki, S., Arikawa, Y. and Toko, K. (1996). *Mater. Sci. Eng.*, C4, 45.
67. Imamura, T., Toko, K., Yanagisawa, S. and Kume, T. (1996). *Sens. Actuators*, B37, 179.
68. Bartoshuk, L. M. (1975). *Physiol. Behav.*, 14, 643.
69. Takagi, S., Toko, K., Wada, K., Yamada, H. and Toyoshima, K. (1998). *J. Pharm. Sci.*, 87, 552.
70. Katsuragi, Y. and Kurihara, K. (1993). *Nature*, 365, 213.
71. Katsuragi, Y., Mitsui, Y., Umeda, T., Otsuji, K., Yamasawa, S. and Kurihara, K. (1997). *Pharm. Res.*, 14, 720.
72. Takagi, S., Toko, K., Wada, K. and Ohki, T. (1999). *Dig. Tech. Papers Transducers '99*, p. 1638.
73. Toko, K., Yasuda, R., Ezaki, S. and Fujiyoshi, T. (1998). *Trans. IEE Jpn*, 118-E, 1.

7

Other methods to measure taste

7.1 Impedance measurement

A multichannel taste sensor using lipid/polymer membranes based on the measurement of electric potential responds to taste in a manner similar to a human gustatory sensation but with better reproducibility and higher resolution. The sensor output is similar for chemical substances producing similar taste qualities and is very different for those producing different types of taste. The taste sensor is effective in discriminating and quantifying the taste of foodstuffs such as coffee, beer, sake, mineral water and milk. The taste of amino acids was also quantified using the multichannel taste sensor. These results may provide an objective scale of human sensory expressions.

Whereas the taste sensor based on potentiometry can detect nonelectrolytes and weak electrolytes, the sensor outputs are smaller than those for strong electrolytes such as NaCl and HCl. This is because the electric potential response in lipid membranes of the taste sensor is dependent on changes of the surface potential produced by the diffuse electric double layer, and nonelectrolytes and weak electrolytes have very little effect on the electric double layer, even if they are bound to the membrane. In general, it is considered that physicochemical interactions such as binding to lipids affect the structure of lipid/polymer membranes. Consequently, it may be possible to detect such interactions by measuring changes in membrane impedance composed of electric resistance and electric capacitance. The studies described below involve measuring changes in the impedance of lipid/polymer membranes by the application of chemical substances which produce a taste.[1]

7.1.1 Thin lipid/polymer membranes and measuring apparatus

The lipid/polymer membranes used in this work consisted of lipids, PVC as the polymer and DOPP as the plasticizer (see p. 115). In this work, three types of lipid, DOP, OAm and TOMA, were used. A 0.01 ml sample from a solution of 80 mg PVC, 4 ml THF, 0.1 ml DOPP and 0.04 ml DOP (or 5 µl OAm or

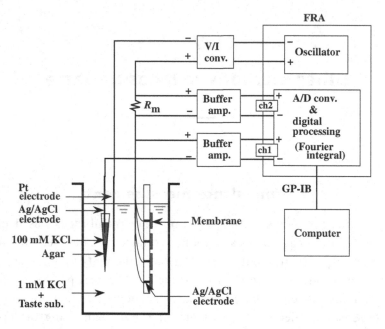

Figure 7.1. Experimental apparatus to measure the impedance of lipid membrane.[1] V/I conv., voltage/current converter; A/D conv., AC–DC converter; FRA, frequency response analyzer; GP–IB, general purpose interface bus.

0.0269 ml TOMA) was applied to coat the Ag/AgCl electrode (see also Fig. 7.1). The width of the resulting membrane was estimated to be about 5 μm, which is one order smaller than that of the usual lipid/polymer membranes used so far.

Figure 7.1 shows the experimental apparatus used to measure the impedance of the lipid/polymer membrane. The impedance of the membrane was evaluated using FRA (frequency response analyzer). The alternating input current was generated by an oscillator and a voltage/current converter and applied across the membrane using a platinum electrode. The input current was measured as the voltage drop across a monitor resistance (R_m) inserted into the input circuit. The output voltage across the membrane was measured using a pair of Ag/AgCl electrodes covered with agar salt. The impedance of the membrane was calculated using the ratio of amplitude and the phase difference between the input current and the output voltage.

7.1.2 Membrane impedance changes caused by taste substances

Figure 7.2 shows changes in the impedances of DOP and TOMA membranes in 1 mM KCl solution plotted on a complex plane at changing

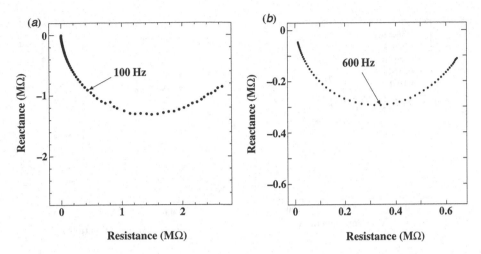

Figure 7.2. Vector traces of the impedance of (*a*) DOP and (*b*) TOMA membranes with input frequency change.[1]

input frequencies. Half circular traces were obtained for these membranes. This result shows that the equivalent circuit of the impedance is approximately represented by the parallel-connected circuit of resistance and capacitance.

Figure 7.3 shows the changes of membrane resistance (ΔR) with increasing concentration of taste substance. The calculation was performed at 100, 1000 and 600 Hz input frequencies for the DOP, OAm and TOMA membranes, respectively. These frequencies were chosen to achieve a low level of noise and a relatively large magnitude of impedance on the complex plane. The initial membrane resistance was about 1.1 MΩ in the DOP membrane and 1.8 MΩ in the OAm membrane, while it was as low as about 400 kΩ in the TOMA membrane. The resistance of the lipid/polymer membranes of DOP and OAm tended to be reduced by all of the strong electrolytes (NaCl, HCl, quinine and MSG), whereas a peak was observed for both membranes at the intermediate concentration of MSG. In a similar way, a peak appeared for NaCl response in the DOP membrane. Sucrose (a sweet nonelectrolyte) increased the resistance of both the membranes.

The thresholds at which the changes appear are about 30 μM for HCl, quinine and MSG in the DOP membrane, while they are about 1 mM for NaCl and sucrose. In the OAm membrane, they are about 30 μM for HCl and MSG, 0.1 mM for quinine, about 0.3 mM for sucrose and NaCl.

The TOMA membrane showed the same decreasing tendency for electrolytes as HCl, MSG and quinine; however, the changes were smaller than those of the DOP and OAm membranes even if the relative changes (%)

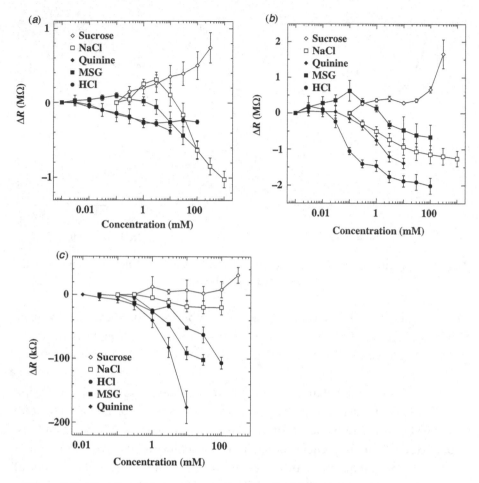

Figure 7.3. Changes of the resistance of various lipid/polymer membranes with taste substances:[1] (a) DOP, (b) OAm and (c) TOMA membranes.

were considered in relation to the initial values. No change occurred with sucrose.

Figure 7.4 shows the changes of membrane capacitance (ΔC) of DOP and TOMA membranes. The initial membrane capacitance was about 1.5, 1.1 and 0.5 nF in the DOP, OAm and TOMA membranes, respectively. It increased significantly for HCl in the DOP membrane, with smaller increases for the other three substances, NaCl, quinine and MSG. However, sucrose produced no effect. In the OAm membrane, the relative change amounted to 10% even for MSG, which caused the largest change among the five chemical substances. The changes were insignificant for the others. MSG caused the largest increase among the electrolytes with the TOMA membrane. No change was seen for the nonelectrolyte sucrose.

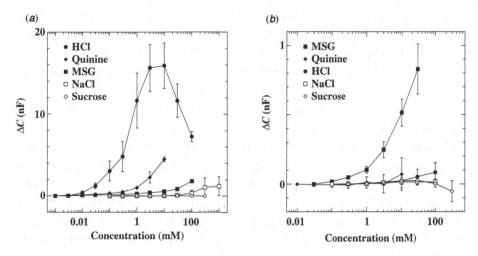

Figure 7.4. Changes of the capacitance of lipid/polymer membranes with taste substances:[1] (*a*) DOP and (*b*) TOMA membranes.

The following discussion mainly concerns the DOP membrane, because it showed large changes in resistance and capacitance. The thresholds, about 30 μM for HCl and 1 mM for NaCl, in the DOP membrane are of an order comparable to those obtained in measurements of membrane potential using the conventional taste sensor and are somewhat superior to results from human sensory evaluations. The response to quinine shows that the threshold of 30 μM is higher than those obtained in the DA and OA membranes of the conventional taste sensor (see Fig. 6.5), about 1 μM, and is comparable to that in sensory evaluations. The threshold for MSG, about 30 μM, is lower than those of the conventional taste sensor and sensory evaluations. The result is significant for sucrose; a large increase appeared from approximately 1 mM, which is much lower than approximately 200 mM detected by humans.

The decreases in resistance of the membranes for NaCl, HCl, quinine and MSG in Fig. 7.3 can be explained as follows: molecules of strong electrolyte play the role of charge carriers within the membrane and the increase in carriers causes an equivalent decrease of resistivity of the membrane. In addition, the adsorption of quinine or MSG onto the membrane[2,3] may increase the hydrophilicity, which results in increasing permeation of ions such as K^+ and Cl^- contained in the solution.

The increase of resistance with sucrose implies that sucrose is adsorbed onto the hydrophilic surface of the membrane and may increase the packing density of the membrane equivalently or block ion permeation. It causes a decrease in ion permeation through the membrane.

These results are very similar to those obtained using cyclic voltammetry in a monolayer lipid (octadecyl mercaptan) membrane.[4] The present method seems to be far superior to the previous one both in terms of ease of membrane preparation and measurement without oxidation–reduction substances. In fact, taste itself cannot be measured, in the strictest sense using a cyclic voltammetric measurement, which requires such oxidation–reduction substances in the taste solution.

The capacitance increased significantly in the DOP membrane, which is negatively charged, with increasing HCl concentration (Fig. 7.4(*a*)). This may be because of increased polarization of the membrane as the outer surface contacting the taste solution decreases its negative charge by proton binding to lipids. In fact, the increase was negligibly small in the OAm membrane for HCl: here Cl^-, which act with positively charged membranes, are scarcely adsorbed.

By comparison, MSG increased the capacitance of the OAm and TOMA membranes. MSG is composed of negatively charged glutamate ions and Na^+. Glutamate anions have been shown to bind to lipid/polymer membranes.[2] In the present case, therefore, it can be considered that the decrease in positive charge of the outer surface of the membrane resulted in increasing polarization, as in the case of the DOP membrane.

Small increases in the capacitance observed at high concentrations of NaCl in the DOP and OAm membranes may be related to the decrease in the width of the electric double layer formed near the membrane surface in an aqueous solution.

The change in resistance was small in the TOMA membrane because it has a low resistance, which implies high permeation of ions, even at low ion concentrations. Although the ion permeation may increase with increasing concentrations of electrolytic taste substances, the rate of increase is low because of its low resistance, as described below. The membrane can be regarded as a parallel connected circuit of many paths through which ions can move.[5,6] As the first approximation, we assume n ion paths in the membrane with resistance r_0. In this case, the membrane resistance R becomes r_0/n. Because the DOP and TOMA membranes have high and low membrane resistance, respectively, r_0 can be regarded as high and low. For simplicity, we assumed n does not differ greatly between these two membranes. Let us consider the simplest situation in which one path permeates ions freely with increasing concentration of electrolytes; the resistance of one path decreases to r', which is low. We can assume $r' \ll r_0$ for the DOP membrane and $r \leq r_0$ for the TOMA membrane, because r_0 should be high and low, respectively, as above. A straightforward calculation shows that $R \approx r'$ as a limiting case for the DOP membrane, whereas it changes slightly from r_0/n for the TOMA membrane. This result agrees with the observed results in Fig. 7.3.

It should be noted that sucrose (nonelectrolyte) could be detected through an increase in the membrane resistance of the DOP and OAm membranes. This result implies that the impedance measurement is effective for the detection of nonelectrolytes. However, the standard deviations show that the errors of repetitive measurement were large; this point must be improved if this method is to be applied to the taste sensing of common foods. If we combine two taste-sensing systems based on different principles, i.e., potentiometric and impedance measurements, a more powerful and realistic taste-sensing system can be realized.

7.2 Surface plasmon resonance

The technique using the phenomenon called surface plasmon resonance (SPR) can make it possible to detect the change of refractivity (or dielectric constant) of the solution in a thin local region (typically about 1 μm thickness), which is very close to the surface of a thin basal metal plate. Interactions between a lipid membrane and taste substances were investigated by using SPR. The lipid membrane utilized as a receptor for taste substances was the LB film composed of dihexadecyl phosphate (DHP).[7]

7.2.1 Principle of SPR measurement with an LB membrane

The surface plasmon is an electron-density wave that is propagating along a surface of a metal facing a dielectric material.[8] By utilizing the resonance phenomenon of the surface plasmon excited by an adequate light beam, it is possible to detect the change of the refractivity of a local region, which is limited within about 1 μm in thickness along a metal film. This method needs neither reporter-bases nor tracers for chemical substances. It is easy to bind various guest molecules as receptors onto the surface. Some studies have been reported using chemical sensors to detect chemical reactions,[9] and biosensors to detect antigen–antibody reactions.[10,11]

In this study, the ATR (attenuated total reflection) method[12,13] was used to excite the surface plasmon. A sample to be measured was placed on a thin basal metal (gold) film that is attached to a glass prism, and the P polarized light beam for exciting the plasmon is introduced into the prism so that the angle of incident is larger than the angle of total reflection, as shown in Fig. 7.5. In this condition, there can exist the evanescent wave near the boundary surface of the metal film facing the glass prism. The resonance

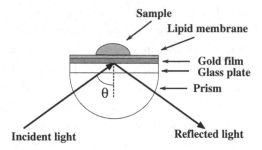

Figure 7.5. Transducer of the surface plasmon resonance sensor system.[7]

phenomenon occurs if a certain condition (which will be mentioned below) holds. In SPR, the energy of the incident light is transmitted to the excited surface plasmon. As a result, the intensity of the reflected light becomes lower. The incident angle, at which the maximum attenuation occurs, is called the resonance angle. The wave number K_{sp} of the surface plasmon, propagating along the boundary surface, can be obtained as a function of its frequency by solving the Maxwell equation, and given by

$$K_{sp} = \frac{\omega}{c} \sqrt{\frac{\epsilon_m(\omega)\epsilon_s}{\epsilon_m(\omega) + \epsilon_s}}, \tag{7.1}$$

where $\epsilon_m(\omega)$ is the complex dielectric constant of the thin gold film, ϵ_s the dielectric constant of a material facing the gold plate, ω an angular frequency of the surface plasmon wave, and c is the velocity of light. The wave number K_E of the evanescent wave ($K_P \sin \theta$), i.e., the projection of the wave number K_P of the incident light beam in the prism, satisfies the condition at the angle θ:

$$K_E = K_P \sin \theta = K_{sp}. \tag{7.2}$$

This means that the resonance occurs by coincidence between the wave number of the evanescent wave and that of the surface plasmon wave. The angle θ that suffices the above equation is the resonance angle, which is dependent on the refractive index near the surface of the gold film facing the sample to be measured.

Figure 7.5 shows the detecting part of the transducer of the measurement system. A glass plate coated with the gold thin film (50 nm thickness) is placed on the prism. This measurement system has a set of two detectors, which are the reference channel and the sample channel. The LB membrane was placed on the sample (S) side but not on the reference (R) side. First, distilled water (10 µl) is dropped on both the R and the S sides; then, the taste solution is added to the waterdrops of both sides to give the same concentration on both sides. The differential measurement using the two channels gives the interaction characteristics between taste substances and the lipid membrane. Here, the LB membrane was fabricated by using the vertical dipping method, and the lipid, DHP, was built up on the gold thin film on the glass plate. The bath solution for fabricating the LB membrane contained ultrapure water, $BaCl_2$ and $NaHCO_3$. The surface pressure of the lipid layer expanded on the solution was kept at 30 mN/m.

The laser beam (wavelength = 670 nm) was introduced from the prism side toward the gold film with an angle larger than the angle of total reflection. The intensity of the reflected light was measured by scanning the incident angle.

7.2.2 Changes of resonance angle with taste substances

The SPR method fundamentally detects the change of a dielectric constant or refractive index of the sample concerned. Initially, the resonance curves of five types of taste substance at various concentrations was examined using only the R side (the gold film with no coating of lipid membrane), in order to investigate the pure effect of the refractivity of the samples on the resonance curve. The result for quinine is shown in Fig. 7.6. It was found that the resonance curve (the resonance angle) shifted toward larger angles with increasing concentration. The resolution of the obtained resonance angle was $\pm 0.1^{\circ}$ because the obtained angles differed by about $\pm 0.1^{\circ}$ under the same conditions.

The resonance angle changed by about 1° with each layer of the LB membrane. The most effective number of layers was, therefore, determined to be six because the effective measurement range is about 10°. The change of

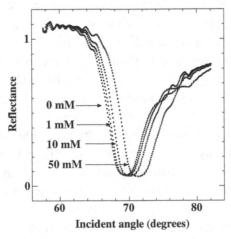

Figure 7.6. Resonance curves for quinine with different concentrations.[7]

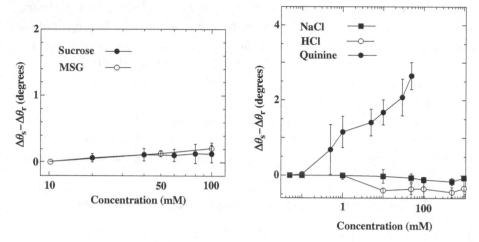

Figure 7.7. Dependences of $\Delta\theta_s - \Delta\theta_r$ of the six-layered LB membrane on five kinds of taste substance.[7]

refractivity near the membrane could be detected with sufficient sensitivity because the thickness of the six-layered LB membrane was 10 nm, while that of the effective measurement range was about 1 μm. Measurements were made 20 min after the sample was added.

The result is shown in Fig. 7.7. The abscissa represents the concentration of taste substances, and the ordinate represents the differences $\Delta\theta_s - \Delta\theta_r$ of the change of the resonance angle in the S side from that in the R side. The difference between the changes of resonance angles is expected to reflect the interactions between the taste substances and the lipid LB membrane. As shown in Fig. 7.7, there was no change in $\Delta\theta_s - \Delta\theta_r$ with sucrose, NaCl, HCl and MSG; the effects were the same on both the S and R sides. However, for quinine there was a change in the resonance angles, which was seen at values above a threshold of 100 μM at which the value on LB membrane side was 10 times smaller than that on the gold film R side.

It was concluded that quinine molecules are adsorbed onto the lipid LB membrane from both the time course of response and from the dependence of the response magnitude on the number of LB layer.

Bitter sensation is suppressed by sweet or salty taste substances in biological systems and this effect may be detectable by the SPR method. Differences in the resonance angles were measured by adding quinine in the presence of NaCl (or sucrose), and the results were compared with quinine plus distilled water.

Figure 7.8 shows the changes in the resonance angles for quinine in the presence of 100 mM NaCl. A higher concentration of quinine was required

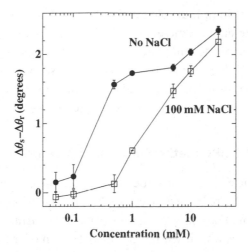

Figure 7.8. The use of surface plasmon resonance to detect suppression of quinine adsorption onto the lipid membrane by NaCl.[7]

to cause a change in resonance angle in the presence of NaCl than without it; this suggests that the adsorption of quinine was suppressed by NaCl. This phenomenon may cause the decreasing intensity of bitterness when Na^+ is present.[14]

The result obtained using sucrose was similar to that using NaCl. From these results, it appears that the SPR method can detect interactions between taste substances, as did measurement of membrane potential (see Section 6.9). This suggests that the suppressive effect seen in the measurement of membrane potential is a consequence of suppression of absorption. This suggestion may also bring a new insight to the interaction mechanism in the suppression effect observed in biological systems.

The multilayered LB film composed of DHP and the sugar-binding protein ConA was also investigated using the SPR method.[15,16] It was found that the LB membrane including ConA shows larger responses to taste substances such as quinine, caffeine and sucrose than the pure LB membrane. This may suggest the importance of proteins in sensor technologies for measuring taste.

7.3 Surface photo-voltage method

A multichannel taste sensor using an SPV (surface photo-voltage) method with lipid membranes has been proposed.[17] Its principle is almost the same as the odor-sensing device using FET mentioned in Section 5.5. The advantage of this method lies in its contactless approach, leading to a simple system that enables integration of multiple sensing elements on a single semiconductor surface. For the enhancement of sensitivity to sweet substances, a new differential measurement technique combined with a base signal suppression method was adopted; hence stabilization of the sensor response was attained by canceling the common noise and drift components by taking the differential signal between the sensing region and the reference region.

7.3.1 SPV method with lipid membranes

Figure 7.9 shows the SPV sensor structure and the measurement system. The electrode consists of a reference electrode/sample solution/lipid membrane/oxide semiconductor. The change of a semiconductor surface potential caused by the taste substances was measured by the detection of the photo-generated current. An alternatively modulated light beam (1 mm ϕ) was irradiated onto the silicon surface and the photo-current generated in the surface depletion layer was picked up by a lock-in amplifier. For the differential mode measurement, two light beams were alternatively irradiated on the different areas (sensing region and reference region) of a semiconductor surface and then differentially superimposed using a phase-sensitive detector.

A sample substrate of an n-type silicon (1–2 Ωcm), covered by an insulator (SiO_2 : 50 nm, Si_3N_4 : 50 nm), was treated by (3-amino-propyl)-triethoxysilane to make the surface hydrophobic. The lipid materials used were OA, L-α-phosphatidylethanolamine (PEA), cholesterol (C), lecithin (L) and the

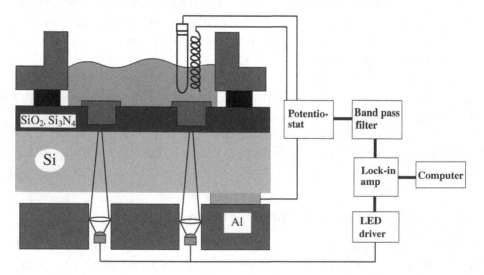

Figure 7.9. Electrode structure and measurement system of a surface photo-voltage sensor.[17]

equimolar mixture of OA and PEA (Mix). They were deposited on the hydrophobic semiconductor surface at different areas.

Lipid materials were cast on the polyvinyl butyral (PVB) membrane coated on the Si_3N_4 surface for immobilization within the PVB membrane.

7.3.2 Responses to five basic tastes

Figure 7.10 shows the response potentials of a cholesterol-modified sensor to the five basic taste substances. Different response magnitudes for different taste substances suggest that a taste sensor could be constructed by analyzing the response patterns of multiple lipid membranes.

In a similar way to the conventional taste sensor mentioned in Chapter 6, the largest response is seen with quinine and the smallest with sucrose. This fact may be common in measurements of electric potential using lipid membranes.

Figure 7.11 shows response patterns of five lipid membranes to various sweet substances using the differential SPV signal. The improvement of signal to noise ratio of the output of Fig. 7.11 is about two orders of magnitude compared with that of Fig. 7.10. It comes from a noise compensation effect through the differential measurement process. Considerable difference of these patterns for sugars suggests the possibility to discriminate the quality of sweet substances.

For the patterning of the sensing film, an improved LB technique based on a horizontal lifting method was also tried. The lipid membranes were

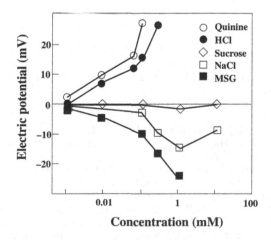

Figure 7.10. Response voltage of cholesterol-modified electrode to five basic taste substances.[17]

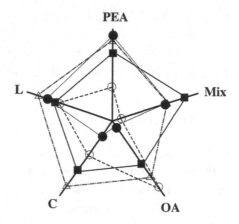

Figure 7.11. Response patterns of five lipid membranes for various sweet substances using differential surface photo-voltage method.[17] Sucrose (○), lactose (●), maltose (△) and lavulose (■).

transferred on the Si_3N_4 surface by making contact to the L (Langmuir) film formed on the top of a droplet of distilled water which was pushed out through a tiny hole (1 mm ϕ) of a Teflon (PTFE) plate. The LB film can be compressed before the deposition by shrinking the volume of the droplet that is attained by moving the Teflon plate upward. This technique made it possible to integrate multiple sensing materials at designed positions by shifting the position of the hole.

This device shows one possible direction for development of a miniaturized, integrated taste sensor. Future studies may produce a stable system suitable for practical use.

REFERENCES

1. Toko, K., Akiyama, H., Chishaki, K., Ezaki, S., Iyota, T. and Yamafuji, K. (1997). *Sens. Mater.*, 9, 321.
2. Oohira, K., Toko, K., Akiyama, H., Yoshihara, H. and Yamafuji, K. (1995). *J. Phys. Soc. Jpn*, 64, 3554.
3. Oohira, K. and Toko, K. (1996). *Biophys. Chem.*, 61, 29.
4. Iiyama, S., Miyazaki, Y., Hayashi, K., Toko, K., Yamafuji, K., Ikezaki, H. and Sato, K. (1992). *Sens. Mater.*, 4, 21.
5. Kobatake, Y., Irimajiri, A. and Matsumoto, N. (1970). *Biophys. J.*, 10, 728.
6. Toko, K., Tsukiji, M., Ezaki, S. and Yamafuji, K. (1984). *Biophys. Chem.*, 20, 39.
7. Yasuda, R., Toko, K., Akiyama, H., Kaneishi, T., Matsuno, T., Ezaki, S. and Yamafuji, K. (1997). *Trans. IEICE Jpn*, J80-C-II, 1 [in Japanese]; *Electronics and Communications in Japan*, Part II, 80, 1 [English translation].
8. Kittel, C. (1996). *Introduction to Solid State Physics*, 7th edn. John Wiley, New York, Ch. 10.
9. Kawata, S. and Minami, S. (1987). *Kougaku (Optics)*, 16, 438 [in Japanese].

10. Sun, X., Matsui, Y., Shiokawa, Y. and Kubota, H. (1991). *Trans. IEICE Jpn*, J74-C-II, 443 [in Japanese].
11. Liedberg, B., Lundström, I. and Stenberg, E. (1993). *Sens. Actuators*, B11, 63.
12. Mirlin, D.N. (1982) In *Surface Polaritons*, Vol. 1, eds. Agranvoic,V.M. & Mills, D.L. North-Holland, Amsterdam, p. 17.
13. Otto, A. (1968). *Z. Phys.*, 216, 398.
14. Breslin, P.A.S. and Beauchamp, G.K. (1997). *Nature*, 387, 563.
15. Adachi, T. and Toko, K. (1998). *Trans. IEE Jpn*, 118-E, 371 [in Japanese].
16. Adachi, T. and Toko, K. (1999). *Dig. Tech. Papers of Transducers '99*, p. 1660.
17. Katsube, T., Sasaki, Y. and Uchida, H. (1995). *Tech. Digest 13th Sens. Symp.*, p. 49.

8

Toward a sensor to reproduce
human senses

―――

8.1 Discrimination of wine flavor using taste and odor sensors

Earlier chapters have reviewed a range of sensor types. Recently, there have been striking developments in odor and taste sensors. As shown in Fig. 8.1, it may not be impossible to discuss deliciousness of foods by combining the sensors corresponding to the five senses. We can use an optical sensor for the color, a thermometer for the temperature, a pressure sensor for the texture, a taste sensor for the taste and an odor sensor for the odor. Analysis of outputs from the above sensors, together with environmental and personal information, may lead to quantification of deliciousness. The following example describes a first step of this trial.

Wine has both taste and odor qualities resulting from different aromatic molecules in the liquid and vapor phases. The average wine contains about 80–85% water and over 500 different substances, some of which are very important to the wine flavor in spite of their low concentrations. The main groups are acids, alcohols, esters, sugars and tannins. The difference in the color of wine comes mainly from tannins (which are also responsible for the flavor). The tannins are present in the solid part of the grapes, which are fermented together with the liquid part when a red wine is made; for white wines only the liquid from grapes is used. It is known that tannins are responsible for the astringency of a solution whereas some acids are responsible for "freshness" and others for peculiar odor nuances.[1]

Among this huge number of molecules, those which can easily pass in a vapor phase represent potential stimulus for the human olfaction when they reach the bulb in the nasal cavity and interact with the odor receptors. Molecules in the liquid phase are responsible for the perceived taste when they interact with taste buds on the tongue. The overall perception of a substance as far as its chemical properties are concerned results from a combination of odor and taste senses and also from the so-called trigeminal sense (responsive to irritant chemical species). This perception is hereafter

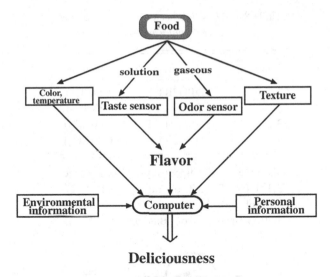

Figure 8.1. Measurement of deliciousness.

referred to as flavor. Wine, therefore, is a suitable candidate for testing the performance of the sensory fusion of taste and odor sensors.[2] The odor-sensor array used in this study is composed of four diffcrent novel conducting polymers that have been recently developed.[3] The monomer (25 mg) is dissolved in trichloroethylene (2 ml) and the oxidizing salt previously dissolved in acetonitrile is added in a dropwise manner. The polymerization process then occurs and the resulting solution is sprayed onto an alumina substrate where four interdigitated electrodes were previously evaporated. After evaporating the solvent, the conducting polymer is connected with the four electrodes, as shown in Fig. 8.2. Four different sensing elements were obtained by combining two different monomers and two oxidizing salts (see Table 8.1).

The electric resistance between the inner electrodes of these sensing elements ranged from 1 to 100 kΩ. The resistance measured at the inner electrodes varied when volatile molecules were adsorbed at the surface of the polymer film. The average sensitivity, expressed as the ratio of the resistance change to the base resistance value, was almost always less than 2% for the elements used in wine sensing. These sensors show broad and overlapped sensitivities to many compounds such as alcohols, amines, hydrocarbons and phenols. They also show a cross-sensitivity to water vapor (relative humidity; RH) so that monitoring of RH was necessary during the experiments.

The taste-sensor array is composed of eight different polymer/lipid membranes, as shown in Table 6.1 (p. 116). After preparation, the electrodes were immersed in a solution of a Japanese red wine for four weeks (the wine used as a standard solution) before they were used for the experiments

Table 8.1. *Materials used for the conducting polymer forming process*

Channel	Monomer	Oxidizer
1	3DPO2BT	$Fe(ClO_4)_3$
2	3,3'-DPTTT	$Fe(ClO_4)_3$
3	3,3'-DPTTT	$Cu(ClO_4)$
4	3DPO2BT	$Cu(ClO_4)$

3DPO2BT, 3,3'-dipentoxy-2,2'-bitiophene; 3,3'-DPTTT, 3,3'-dipentoxy-2,2': 5',2''-tertiophene.
With permission from Baldacci *et al.*[2]

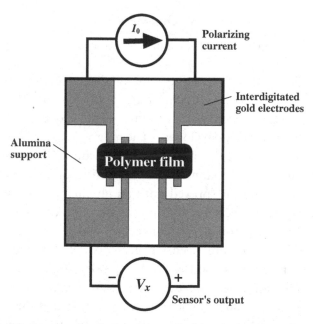

Figure 8.2. Layout of a single odor-sensing element on alumina substrate. (With permission from Baldacci *et al.*[2])

(preconditioning method). The four different wines used for the experiments are listed in Table 8.2.

The raw data were normalized by the following method. Let $\overline{S}_{is}$ be the set of measurements taken with the multichannel odor sensor:

$$\overline{S}_{is} = \begin{bmatrix} S_{11} & S_{21} & S_{31} & S_{41} \\ \vdots & & & \vdots \\ \vdots & & & \vdots \\ S_{1N} & \cdots & \cdots & S_{4N} \end{bmatrix}, \tag{8.1}$$

Table 8.2. *Four different wines used in the experiments*

Wine	Brand name
Wine 1 (white)	Est! Est! Est! di Montefiascone 1995, Italy
Wine 2 (red) (standard solution)	Bon Marche' Mercian, Japan
Wine 3 (white)	Chablis 1994, France
Wine 4 (red)	Rosso di Montalcino, Fattoria dei Barbi, 1994, Italy

where i is the channel number, s is the s-th measurement and N is the number of measurements. Each element of the matrix represents the response of a single element of the odor-sensor array expressed as the ratio of the maximum resistance change, upon exposure to the vapors, to the base resistance measured in presence of clean air flowing through the exposition chamber. Let $\overline{T}_{is}$ be the set of measurements taken with the multichannel taste sensor:

$$\overline{T}_{is} = \begin{bmatrix} T_{11} & \cdots & \cdots & T_{81} \\ \vdots & & & \vdots \\ \vdots & & & \vdots \\ T_{1N} & \cdots & \cdots & T_{8N} \end{bmatrix}, \tag{8.2}$$

where i, s and N have the same meanings as above. Each element of the matrix represents the response of a single electrode of the taste sensor expressed as the difference between the electric potential in the testing solution and the electric potential in the standard solution.

The mean values of the responses for each channel are

$$\overline{S}_i = \frac{1}{N} \sum_{s=1}^{N} \overline{S}_{is} \tag{8.3a}$$

for the odor sensor, and

$$\overline{T}_i = \frac{1}{N} \sum_{s=1}^{N} \overline{T}_{is} \tag{8.3b}$$

for the taste sensor.

Computing the average of the square errors of the responses first among the samples and then among the channels we obtain

$$\sigma_o^2 = \frac{1}{4N} \sum_{i=1}^{4} \sum_{s=1}^{N} (\overline{S}_{is} - \overline{S}_i)^2 \tag{8.4a}$$

for the odor sensor and

$$\sigma_t^2 = \frac{1}{8N} \sum_{i=1}^{8} \sum_{s=1}^{N} (\overline{T}_{is} - \overline{T}_i)^2 \qquad (8.4b)$$

for the taste sensor.

Now the original sets of data can be normalized as follows:

$$\overline{s}_{is} = \frac{\overline{S}_{is} - \overline{S}_i}{\sigma_o} \qquad (8.5a)$$

for the odor sensor, and

$$\overline{t}_{is} = \frac{\overline{T}_{is} - \overline{T}_i}{\sigma_t} \qquad (8.5b)$$

for the taste sensor. Combining the normalized sets of data we obtain the data set:

$$\overline{F}_{is} = [\overline{s}_{is} \quad \overline{t}_{is}], \qquad (8.6)$$

that is, a $12 \times N$ dimensional data array.

From each measurement using the taste sensor, an eight-dimensional vector representing eight membrane potentials was extracted. One cycle of measurements consisted of four different acquisitions made by rotating the testing samples in the following order: wine 1, wine 2, wine 3 and wine 4 (note here that wine 2 was also used as the standard solution). PCA was made after normalizing. The distribution of data in the principal component space is shown in Fig. 8.3.

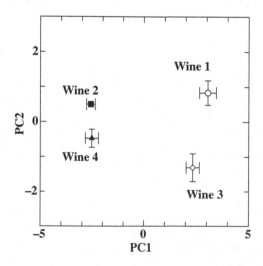

Figure 8.3. Results of the PCA applied to the data set from the taste sensor. Wines 1 and 3 are white; wines 2 and 4 are red. (With permission from Baldacci *et al.*[2])

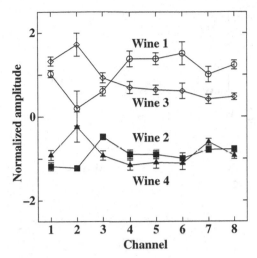

Figure 8.4. Eight-dimensional taste patterns of the testing wines.
(With permission from Baldacci *et al.*[2])

We can observe that the data are clustered in four well-separated groups representing the four different wines used. The PC1 accounts for the differences between red and white wines, whereas the PC2 accounts for the differences between wines of the same color. As mentioned above, the color of wine is mainly from the content of tannins. The sensitivity of the taste sensor to tannins was investigated in a previous work[4] to show that the array's element, DOP:TOMA 3:7, was the most sensitive for tannic acid. This agreed with the present result because the element DOP:TOMA 3:7 is the larger contributor to the PC1, which, in turn, accounts for the discrimination between red and white wines.

Figure 8.4 shows the averaged taste patterns of the four wines. The taste pattern can be seen as the fingerprint of a wine in the eight-dimensional space represented by the taste sensor. Each wine is characterized by its own taste pattern.

From each measurement using the odor sensor, a four-dimensional data vector representing the peak values of the electric resistance changes was extracted. The PCA was computed after normalizing the data in the same way as for the taste sensor. The plot in Fig. 8.5 shows the distribution of clusters in the principal component space. Again, discrimination among different wines was achieved. At this stage no data were available concerning the sensitivity of odor sensors to the main components of wines and therefore no quantitative consideration can be put forward to account for these results. Nevertheless, in this case the mutual distribution of clusters differs from that in Fig. 8.3 and this can be considered as evidence that the information concerning the samples

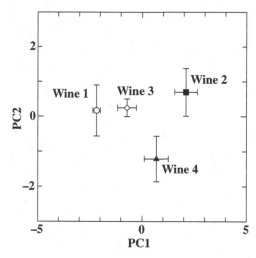

Figure 8.5. Results of the PCA applied to the data set from the odor sensor.
(With permission from Baldacci *et al.*[2])

provided by the odor sensors accounts for different characteristics of the wines themselves. In this case, the PC1 is still responsible for the discrimination between red and white wines and also for the differences between the two white wines and the two red wines. The PC2 gives information about further differences between the two red wines. Note here, in the PCA plot, how the overall information about the set of samples is topologically distributed in a different way compared with that from the taste sensor.

Figure 8.6 shows the averaged odor patterns for the four wines tested. Again, the odor patterns are the representation of the wines in

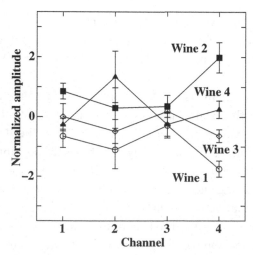

Figure 8.6. Four-dimensional odor patterns of the testing wines.
(With permission from Baldacci *et al.*[2])

the four-dimensional space of the odor-sensing elements. Each sample has its own odor pattern, which differs slightly among samples belonging to the same kind of wine because the responses of the odor sensor and the fluid-dynamic conditions[5] were not highly reproducible, as in the case of the taste sensor.

After either set of data was normalized, according to the method described, a 12-dimensional data array was obtained for each measurement. The 12-dimensional data array is composed of the four-dimensional data array of the odor sensor and the eight-dimensional data array of the taste sensor. Another PCA was performed on this new set of data and the results are shown in Fig. 8.7.

The relative positioning of the clusters in the principal component plane was similar to that for the odor sensor, and the relatively large distance between clusters of the red (2 and 4) and white (1 and 3) wines were successfully achieved by the contribution of the taste sensor, as in Fig. 8.3. The combination of the two sets of data has led to a new representation of the samples in the 12-dimensional space, which we refer to as flavor pattern. A flavor pattern simultaneously contains information from odor and taste sensors concerning the sample measured.

The feasibility of a sensory fusion between taste and odor sensors has been investigated with the aim of discriminating among substances with subtle differences, such as the wines used here. The same method was applied to detect the change in flavor after opening wine, using five samples of the same

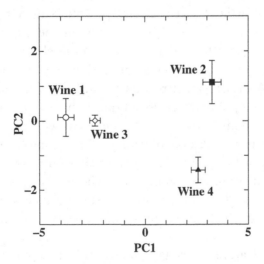

Figure 8.7. Results of the PCA applied to the combination of the data set from the odor sensor with the data set from the taste sensor. (With permission from Baldacci *et al.*[2])

wine opened for different times.[2] Discrimination of wines, having the same denomination but coming from different vineyards, was successfully made using an array of metal oxide semiconductor gas sensors.[6] The sensor fusion is very effective because the information provided by each array is to some extent independent from the others; the arrays account for different characteristics of the wines themselves since the relative positioning of clusters in the principal component space differs from one wine to the next. Conventional multiple sensor arrays have several sensing elements produced by similar technology, e.g., conducting polymer sensor, metal oxide sensor and lipid-membrane sensor. These sensors show broad sensitivities to certain groups of substances but are not sensitive to other compounds. If different types of sensor technology are simultaneously applied, provided that the data from the different sources are independent, it is worthwhile to combine them to obtain a broader viewpoint of the samples measured.

8.2 Perspective

A multichannel taste sensor, i.e., electronic tongue, utilizes lipid membranes as the sensing part. This sensor can discriminate and quantify the taste of chemical substances that is really felt by humans. A recent study[7] shows that measurement of the electric potential of the membrane electrode in standard KCl solution without rinsing the electrode, onto which chemical substances such as quinine and tannin were adsorbed by measuring the sample solution, is very effective for quantifying the taste. This measurement is called a CPA measurement, because the change of electric potential caused by adsorption of chemical substances is measured. The CPA measurement provides information that is different from the usual measurement of electric potential of the sample (i.e., the response electric potential) as described in Chapter 6. Briefly speaking, the CPA measurement reflects the adsorbed amount of chemical substances on the lipid/polymer membranes mainly from hydrophobic interactions. If we combine the CPA measurement with the usual measurement, which detects the electric potential of the sample, we can obtain more information of the overall taste comprising salty, sour, sweet, bitter, umami taste and astringency components. The overall taste of green tea, where amino acids and tannin are main taste substances, was quantified using this method, and high correlations with the results of sensory tests were found.[8]

One of the main features of the taste sensor utilizing lipid membranes is its ability to quantify the taste of amino acids, as seen in Section 6.4. The taste sensor can classify the taste of amino acids into tastes such as bitter, sweet, sour and umami taste. The mixed taste composed of bitterness and sweetness shown by L-methionine and L-valine can be reproduced using the taste sensor by a mixture of L-alanine, which tastes sweet, and L-tryptophan, which tastes bitter. The bitter taste of amino acids is intimately related to the hydrophobicity. The taste of dipeptides can also be discriminated by the response pattern of the taste sensor, which is characteristic of each taste quality. Figure 8.8 shows a taste map of five taste qualities elicited by typical chemical substances, amino acids and dipeptides (K. Toko and T. Nagamori, unpublished data).

The taste of seafoods such as abalone, sea-urchin, crab, scallop and short-necked clam is determined by the content of amino acids.[9] For example, the taste of sea-urchin can be expressed by L-glutamate, glycine, L-alanine, L-valine and L-methionine at 103, 842, 261, 154 and 47 mg/100 g and an adequate quantity of NaCl. The aqueous solution of this mixture was measured using the taste sensor, and the result was compared with the measurements for sea-urchin, which was crushed and homogenized with a mixer. The correlation coefficient was 0.995 for one of the sea-urchins tested (S. Takagi and K. Toko, unpublished data). There was a tendency for the correlation to become higher for expensive sea-urchins; it is related to the quantity of NaCl compared with that of amino acids. As the sea-urchin is more expensive, the weight of amino acids playing a role in producing the main taste becomes larger.

As mentioned in Chapter 2, the reception mechanism of gustatory systems is as yet unclear. However, the taste sensor using lipid membranes can reproduce the taste experienced by humans in most cases except for sweetness elicited by sugars, as detailed in the last two chapters. The response patterns for five taste qualities differ, and hence these tastes are separated well on the three-dimensional taste map (see Fig. 6.7, p. 122). The response patterns for bitter amino acids such as L-tryptophan resemble that for quinine, which is a typical bitter substance belonging to the alkaloids. The detected threshold is almost the same in both the taste sensor and the gustatory system for each taste quality. Suppression of bitterness by sweet substances and phospholipids can be reproduced well using the taste sensor. The tastes of many foodstuffs such as beer, coffee, milk, mineral water, sake and tomatoes can be quantified and the sensor outputs can agree with sensory tests by humans. How should we interpret these facts?

It may be reasonable to consider that in real systems the lipid membrane plays a role in the reception of bitter substances such as quinine and amino acids, sour substances such as acetic acid and citric acid, and salty substances such as NaCl, KCl and $FeCl_3$. The response patterns are large for these taste

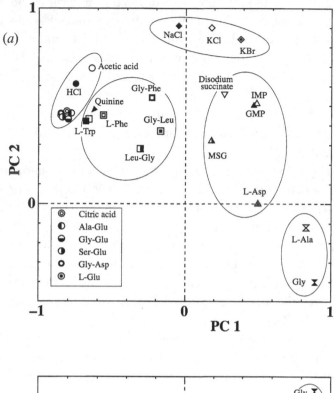

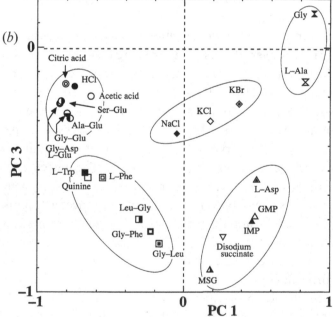

Figure 8.8. Taste map comprising typical chemical substances such as HCl, NaCl, quinine, amino acids and dipeptides. (*a*) PC 1–PC 2, (*b*) PC 1–PC 3 plane.

qualities and differ from one taste quality to another; similar patterns are obtained for chemical substances producing the same taste. Within the group of sweet substances, sugars can be hardly detected using the multi-channel potentiometric taste sensor that uses lipid membranes; however, amino acids (glycine, L-alanine) and artificial sweetners can be measured. Interestingly, a different reception and transduction mechanism in gustatory systems is proposed for these two species, as mentioned in Section 2.3.

The fifth taste quality, umami taste, can be detected using the taste sensor. Three umami taste substances (MSG, IMP and GMP) with very different chemical structures (MSG is an amino acid; IMP and GMP are nucleotides) show almost the same response electric patterns, as seen in Fig. 6.6 (p. 121). There is a possibility that umami taste substances are received at the lipid membrane part of biomembranes, because a similar response pattern can be obtained for other umami substances such as mono-sodium L-aspartate.[10] It can be considered that umami taste is related to chemical substances that have a common structure as neutral salts of weak acids with moderate hydrophobic properties.[11] The hydrophilic group and hydrophobic chains of the lipid membrane interact with the chemical substances that have this structure. A well-known phenomenon that saltiness of NaCl is decreased by coexistent umami taste substances can be reproduced using the taste sensor.[12] However, the reproduction of the synergistic effect experienced by humans is not completely achieved by the taste sensor because the increase in umami taste strength is weaker for the taste sensor than for humans.

Kurihara suggested the importance of the hydrophobic part of lipid membranes for the reception of bitter substances.[13] The results of the taste sensor studies using lipid membranes support this suggestion and furthermore suggest that other taste qualities such as sourness and saltiness can also be detected by the lipid membrane, which can use both its hydrophobic and its hydrophilic parts in reception of chemical stimuli. Living organisms must check the safety of all chemical substances quickly before eating them and it may well be that they use the lipid membrane as well as the proteins embedded in it to achieve this.

Although lipid molecules have a more simple structure than most proteins, they can self-organize into macroscopic multimolecular structures such as the lipid bilayer and liposomes. Once such a structure is formed, it can function as a barrier to ions, a supporting/modulating material for proteins, an electron-transfer medium and as a recognition system at the cell surface.

The studies using the taste sensor described in this book and the results achieved may contribute to our understanding of the reception mechanisms involved in the gustatory system and add new light to the role of lipid membranes.

REFERENCES

1. Boidron, J.N., Lefebvre, A., Riboulet, J.M. and Ribereau-Gayon, P. (1984). *Sci. Aliments*, 4, 609.
2. Baldacci, S., Matsuno, T., Toko, K., Stella, R. and De Rossi, D. (1998). *Sens. Mater.*, 10, 185.
3. Stussi, E., Cella, S., Serra, G. and De Rossi, D. (1996). *Mater. Sci. Eng.*, C4, 27.
4. Iiyama, S., Toko, K., Matsuno, T. and Yamafuji, K. (1994). *Chem. Senses*, 19, 87.
5. Nakamoto, T., Fukuda, A. and Moriizumi, T. (1991). *Sens. Actuators*, B3, 221.
6. Di Natale, C., Davide, F.A.M., D'Amico, A., Nelli, P., Groppelli, S. and Sberveglieri, G. (1996). *Sens. Actuators*, B33, 83.
7. Ikezaki, H., Taniguchi, A. and Toko, K. (1998). *Trans. IEE Jpn*, 118-E, 506 [in Japanese].
8. Ikezaki, H., Taniguchi, A. and Toko, K. (1997). *Trans. IEE Jpn*, 117-E, 465 [in Japanese].
9. Fuke, S. and Konosu, S. (1991). *Physiol. Behav.*, 49, 863.
10. Iiyama, S., Iida, Y. and Toko, K. (1998). *Sens. Mater.*, 10, 475.
11. Hayashi, K., Shimoda, H., Matsufuji, S. and Toko, K. (1999). *Trans. IEE Jpn*, 119-E, 374 [in Japanese].
12. Nagamori, T., Toko, K., Kikkawa, Y., Watanabe, T. and Endou, K. (1999). *Sens. Mater.*, 11, 469.
13. Koyama, N. and Kurihara, K. (1972). *Biochim. Biophys. Acta*, 288, 22.

INDEX